Mathematics Review Workbook for College Physics

H. T. Hudson
University of Houston

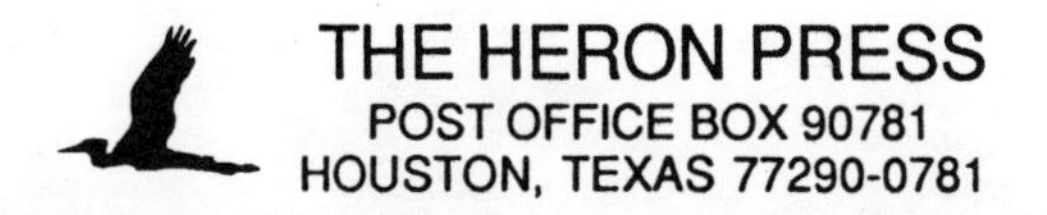

THE HERON PRESS
POST OFFICE BOX 90781
HOUSTON, TEXAS 77290-0781

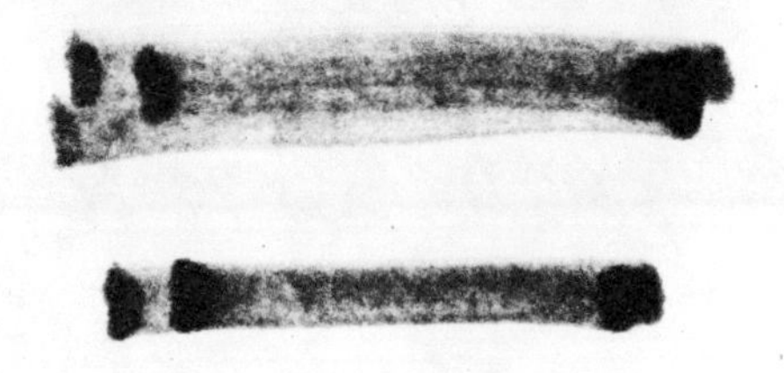

Library of Congress Catalog Card No. 85-23972

ISBN 0-316-37906-9

Second Printing December 1990

Preface

This workbook provides a review of the essential mathematics required for a general survey course in physics. Each chapter addresses a single topic. In so far as possible, each chapter stands alone in that the material concentrates on a narrow range of mathematics skills.

The topics are chosen based on the mathematics needed for a traditional general physics course requiring algebra and trigonometry. If you are enrolling in a physics course that requires calculus, this book should be supplemented with a review of calculus from some other source.

Physics courses require students to be proficient in mathematics because the examinations contain a number of problems requiring the use of mathematics. Knowing how to carry out a mathematical operation is not sufficient. There will be time constraints during examinations. You must be able to use physics to set up a problem from information given, and then solve the equations quickly. It will be of small comfort to know that you could have solved all the problems on a physics test if you fail the test because your take too much time doing mathematics.

The philosophy behind this workbook is that proficiency follows repetition and practice. To that end, there is a minimal amount of explanation followed by examples followed by practice problems. The answers to the practice problems are at the end of the book.

It would be unrealistic to pretend that all students need to review the same mathematics. You may well be able to skip over certain topics. To help you with that decision, the author has developed a diagnostic test that can help you identify the topics you need to review. This test is available free to your physics instructor from the publisher at the address given on the title page.

Physics is considered a difficult course by many students. The three predominant factors that make physics hard are prior misconceptions, analytical reasoning abilities and a knowledge of required mathematics. (Interestingly, only mathematics is listed as prerequisite to physics courses.) This book can only help you with one of the three factors. Don't consider that you have in your hands the magic bullet that will make physics simple. Use the book as intended and you should be able to focus your energies on the physics.

Contents

1 Significant Figures

With the exception of a very few quantities that are defined and not based on measurement, the numbers that occur in physics are never exact. This means that when you do a calculation, the number that you end up with is accurate to only a limited degree. The precision or accuracy of a number is contained in the number itself by the means of significant figures. The significant figures of a number are the digits that are known with certainty plus one digit that is uncertain. Expressing a calculated number to the correct significant figures is called rounding, also known as round-off.

The importance of significant figures is more than cosmetic. In any data set, every value given is accurate to only some maximum level. For example, the device used to measure weight in a laboratory might be a balance of some type. When something is weighed, it is compared to a standard by bringing the balance into equilibrium by adjusting standard weights or a mechanism inside the instrument. If such a scale is brought into balance, the addition of a very small piece of matter, such as a grain of dust, will not affect the reading. The scale is insensitive to that amount of weight. There is an inexactness in the weight by the amount the scale does not detect. In other words, the weight indicated on the scale is only accurate to the range of the least amount of additional weight that will cause the reading to change. Any calculation that implies an accuracy greater than that of the scale might produce a number, but the number will be meaningless. Accuracy cannot be improved by manipulation of the measured numbers, and the result of a calculation should not imply more significance than the least significant number that entered into the calculation.

For example, suppose a city is said to have a population of approximately 230,000. This would mean that the population is closer to 230,000 than to 220,000 or 240,000. Populations of 227,000 or 232,000 would both be catagorized as "approximately 230,000." The population estimate of the city is only accurate to two significant figures, where the count of significant figures starts at the first nonzero digit from the right and goes left to the last digit. (With decimals, the last digit will be the one preceding a string of zeros, as discussed below.)

Now, suppose an accurate headcount is done in the communities that make up the suburbs of this city and it is found that 123,212 people live in these communities. This is an exact number (until someone moves in or out), accurate to six significant figures. The total population of the city (approximately 230,000) plus the suburbs (exactly 123,212) can only be stated to the accuracy of the approximate population of the city. Correctly stated, the total population of city plus suburbs is 350,000. The inaccuracy of the city population makes any implied accuracy for the sum beyond the second digit meaningless.

Most often the word "approximately" is omitted when a number is quoted, with the implication that the number of significant figures is the number of digits up to the string of zeros. If the digit is not significant, it is not stated. In the example just given, a statement that the city has 230,000 people is also a statement that the population is accurately known to only two significant figures.

Of course, this can lead to ambiguity. Zero as a number will occur in counting, and the number of thousands could just very well have been 230 (to three significant figures where the last digit was zero). Since a zero is ambiguous when it appears to the left of the decimal point, the last significant digit may be identified by underlining if it is zero. The fact that a city's population is known to be 230,000 to three significant figures would be indicated by writing the number 23$\underline{0}$,000. If the population were exactly 230,000, this would be indicated by writing the number as 230,00$\underline{0}$.

For decimals, the ambiguity does not arise. Zeros between the decimal point and the first nonzero digit are place holders, and do not enter into the number of significant figures. The number 0.000345 is significant to three figures, just as is the number 456 or 456,000. If a zero occurs at the right end of a decimal, it is always significant (otherwise it would be omitted). The number 3.28900 is significant to six figures.

The concept of rounding is fundamental to significant figures. Suppose the two numbers 2.1 and 3.7 are multiplied on a calculator to produce the product 7.77. The 7.77 has three digits though each of the original numbers has only two. The product should reflect that it is accurate to only two places, so the 7.77 is rounded to 7.8.

> Rule: When the digits following the last significant digit are less than 500...., the last significant figure is unchanged. When the digits following the last significant digit are larger than 500..., the last significant figure is raised by one. When the digits following the last significant digit are 500....(exactly 5), the last significant figure is rounded to the nearest even digit.

(It should be noted that is is not the only rule around. The simple computer algorythms add 5 to the digit following the last significant digit and take the resulting whole number in the last significant digit.) This procedure given in the above rule helps reduce the accumulation of errors.

Under the rule, the number 23.4531 would be rounded to three significant figures as

23.5. The number 23.4999 would be rounded to 23.4. Your instructor or text may use a rule different from the one given here. You should follow whatever rule is appropriate for your course. Given that the last significant digit has some uncertainty, lengthy arguements about the merits of different roundoff methods aren't very productive.

Practice Problems

Round off the following numbers to three significant figures.

1. 6.21845
2. 0.61254
3. 125,499.9
4. 246.15
5. 0.0031461
6. 26.1543

Perform the indicated operation and express the answer to the correct number of significant figures.

7. 2.1 + 0.1561
8. (2.1)(0.1561)
9. 61.5 - 13.331
10. 6/47
11. 6.0/47
12. 0.314 + 0.006312

Answers to Practice Problems

1. 6.22
2. 0.613
3. 125,000
4. 246
5. 0.00315
6. 26.2
7. 2.3
8. 0.33
9. 48.2
10. 0.1
11. 0.13
12. 0.320

Exercises

Carry out the indicated operation and express the answer to the correct number of significant figures.

1. 1/3 = ____________________
2. 1/(3.3) = ____________________
3. 0.4546 + 2.31 = ____________________
4. 1.34528 - 2 = ____________________

5. 0.299 + 1.0 = ____________

6. 6•4 = ____________

7. (1.2)•(222) = ____________

8. 16.2 + (12/3) = ____________

9. $60\underline{0}{,}000 + 13$ = ____________

10. (44)•(2.1356) = ____________

11. 17.2 + 0.00399 = ____________

2

Powers of ten notation

2.1 EXPRESSION IN POWERS OF 10

Physics deals with very small and very large numbers. The most convenient method of expressing such numbers is using powers of 10. Variations on how the powers of 10 are given are called scientific notation or engineering notation. Scientific notation places a decimal after the first digit with the appropriate power of ten. Engineering notation puts the decimal before the first digit. For example, light will travel 9,460,000,000,000,000 meters in one year. This distance is called a light-year and is useful for a number of applications. In scientific notation, the light year would be written as 9.46×10^{15} m. In engineering notation the light-year is written 0.946×10^{16} m.

To convert a number greater than ten to powers of 10 notation, start at the decimal point (or where the decimal is understood to belong) and count to the left. The number of digits counted is the exponent in the ten multiplier or the power of ten. The light year was given as 9,460,000,000,000,000 m. The decimal is understood to follow the last zero, so the number may be written

9,460,000,000,000,000.

Starting at the decimal, counting to the left, there are 13 zeros. Thus

$$9,460,000,000,000,000 = 946 \times 10^{13} \text{ m}.$$

(It will be shown later that the different expressions for the light year, scientific notation, engineering notation or 946×10^{13} m are all the same.)

Other examples are:

$100 = 100. = 1 \times 10^2 = 10^2$

$20{,}000 = 20{,}000. = 2 \times 10^4$

$300{,}000{,}000 = 3 \times 10^8$

$450{,}000 = 45 \times 10^4$

Small numbers are written in powers of 10 notation by utilizing negative exponents. The negative exponent means the reciprocal of the power indicated. For example,

$10^{-1} = 1/10 = 0.1$

$10^{-2} = 1/10^2 = 1/100 = 0.01$

$10^{-6} = 1/10^6 = 1/1{,}000{,}000 = 0.000001$ etc.

The correct power of 10 may be deduced by starting at the decimal point and counting to the right to the new point where the decimal will be located. The number of spaces is the (negative) power of 10.

$0.0002 = 2 \times 10^{-4}$

$0.0000003 = 3 \times 10^{-7}$

$0.000052 = 52 \times 10^{-6}$

For any number, the decimal point may be moved to the right or to the left provided the power of 10 is adjusted to maintain the equality.

Rule: If the decimal is moved to the left, the power of 10 is increased by the number of integers moved and if the decimal is moved to the right, the power of 10 is decreased by the number of integers moved.

$0.13 = 1.3 \times 10^{-1} = 13 \times 10^{-2}$

$4736 = 473.6 \times 10^1 = 47.36 \times 10^2 = 4.736 \times 10^3$

$3.92 = 0.392 \times 10^1 = 392 \times 10^{-2}$

$52 \times 10^{-6} = 5.2 \times 10^{-5}$

Practice Problems

1. $10{,}450 = 1.0450 \times 10^?$

2. $341.24 = 3.4124 \times 10^?$

3. $2{,}500{,}000 = 25 \times 10^?$

4. $7{,}600 = 7.6 \times 10^{?}$

5. $0.000341 = 3.41 \times 10^{?}$

6. $0.000000218 = 21.8 \times 10^{?}$

7. $0.0000262 = 262 \times 10^{?}$

8. $0.00984 = 9.84 \times 10^{?}$

9. 472×10^{-5} = 0.??????

10. 0.58×10^{-4} = 0.???????

11. 1.92×10^{-3} = 0.?????

12. 0.072×10^{-4} = 0.??????

2.2 MULTIPLICATION

Multiplication of numbers expressed as powers of 10 is accomplished by multiplying the integers and adding the exponents.

$$(2 \times 10^{2}) \times (3 \times 10^{7}) = (2 \times 3) \times (10^{2} \times 10^{7}) = 6 \times 10^{2+7} = 6 \times 10^{9}$$

$$(5.2 \times 10^{4}) \times (1.3 \times 10^{5}) = 6.76 \times 10^{4+5} = 6.8 \times 10^{9}$$

$$(1.1 \times 10^{6}) \times (1.6 \times 10^{-19}) = (1.76 \times 10^{6-19}) = (1.8 \times 10^{-13})$$

$$(3.4 \times 10^{-12}) \times (1.8 \times 10^{-11}) = 6.1 \times 10^{-23}$$

Practice Problems

13. $(4 \times 10^{5}) \times (3 \times 10^{2}) =$ ____________

14. $(12 \times 10^{4}) \times (3 \times 10^{7}) =$ ____________

15. $(7 \times 10^{18}) \times (3 \times 10^{3}) =$ ____________

16. $(3 \times 10^{4}) \times (2 \times 10^{6}) =$ ____________

17. $(3 \times 10^{-2}) \times (5 \times 10^{-3}) =$ ____________

18. $(9 \times 10^{-4}) \times (3 \times 10^{-9}) =$ ____________

19. $(4 \times 10^{-1}) \times (6 \times 10^{-5}) =$ ____________

20. $(2 \times 10^{-3}) \times (4 \times 10^{-5}) =$ ____________

21. $(6 \times 10^{-2}) \times (4 \times 10^{6}) =$ ____________

22. $(2 \times 10^3) \times (9 \times 10^{-9}) =$ ________

23. $(8 \times 10^4) \times (3 \times 10^{-4}) =$ ________

24. $(7 \times 10^{-6}) \times (5 \times 10^{10}) =$ ________

2.3 DIVISION

Division of numbers expressed as powers of 10 is accomplished by dividing the integers and subtracting the 10's exponent in the denominator from the 10's exponent in the numerator.

$$\frac{(9 \times 10^6)}{(3 \times 10^4)} = \frac{9}{3} \times 10^{6-4} = 3 \times 10^2$$

$$\frac{19 \times 10^{12}}{28 \times 10^6} = 0.679 \times 10^{12-6} = 0.68 \times 10^6 = 6.8 \times 10^5$$

$$\frac{2 \times 10^4}{6 \times 10^{10}} = 0.33 \times 10^{4-10} = 0.3 \times 10^{-6} = 3 \times 10^{-7}$$

$$\frac{8 \times 10^3}{4 \times 10^{-8}} = 2 \times 10^{3-(-8)} = 2 \times 10^{11}$$

$$\frac{16 \times 10^{-16}}{2 \times 10^{-13}} = 8 \times 10^{-3}$$

Practice Problems

25. $3 \times 10^{-2} \div 6 \times 10^{-4} =$ ________

26. $2 \times 10^4 \div 4 \times 10^{-6} =$ ________

27. $1.4 \times 10^4 \div 7 \times 10^{-2} =$ ________

28. $24 \times 10^9 \div 6 \times 10^{-3} =$ ________

29. $(4 \times 10^{-4})/(2 \times 10^{2}) =$ ____________

30. $(18 \times 10^{4})/(6 \times 10^{-2}) =$ ____________

31. $\dfrac{42 \times 10^{6}}{7 \times 10^{-6}} =$ ______________________

32. $\dfrac{36 \times 10^{-9}}{12 \times 10^{-12}} =$ ______________________

2.4 ADDITION AND SUBTRACTION

In order for numbers to be added or subtracted, they must have the same power of 10. Addition or subtraction does not change that (common) power of 10. For example, the addition

$$(2 \times 10^{3}) + (3 \times 10^{3})$$

gives

$$5 \times 10^{3}.$$

If powers of 10 are different, as in

$$1.5 \times 10^{3} + 2.6 \times 10^{4},$$

the powers of 10 must first be made equal. This may be done as

$$0.15 \times 10^{4} + 2.6 \times 10^{4}$$

or

$$1.5 \times 10^{3} + 26 \times 10^{3}.$$

The sum is either 27.5×10^{3} or 2.75×10^{4}, which round off to significant figures as

$$28 \times 10^{3} \text{ or } 2.8 \times 10^{4}.$$

Practice Problems

33. $1.67 \times 10^{5} + 26.5 \times 10^{3} =$ ____________________

34. $100.2 \times 10^{8} + 99.9 \times 10^{6} =$ ____________________

35. $1.5 \times 10^{-5} + 7.1 \times 10^{-7} =$ ____________________

36. $222 \times 10^{-5} + 4.0 \times 10^{-3} =$ ____________________

37. $1.8567 \times 10^3 - 241.3 \times 10^{-2} =$ ______________

38. $16 \times 10^5 - 42.1 \times 10^2 =$ ______________

39. $7.98421 \times 10^5 - 10^4 =$ ______________

40. $6.4 \times 10^{-17} - 1.6 \times 10^{-19} =$ ______________

2.5 CALCULATORS AND POWERS OF 10

You may key powers of ten directly into a scientific calculator using a key labeled "EE." Suppose you wish to multiply 3.645×10^5 times 2.412×10^3. Key the 3.645, then "EE." Your calculator display will read

3.645 00.

The two zeros off to the right represent the exponents. Key the 5, and the display will read

3.645 05.

Multiplication then proceeds as when exponents are not used. Key the "x", then 2.412 "EE" 3. Finally, key "=" for the answer (8.7917 EE 08).

Negative exponents are entered by changing the sign of a positive exponent with the ± key. The number 2.142×10^{-5} is entered:

2.142 "EE" 5 "±".

The final display reads

2.142 - 5.

Some calcuators require the first entry to be a 1 for direct powers of 10, such as 10^5 or 10^{-11}. If you key "EE" 6, the display will read

0 06.

Pressing "=" gives

0 00.

The calculator has raised zero to the sixth power, which is of course zero.

To enter 10^5, key

1 "EE" 5

and the display reads

1   05.

To enter negative powers of 10, such as 10^{-11}, key

1 "EE" 11 "±".

The display will read

1 - 11.

It is important to remember to key the 1 before the exponent for a direct power of 10. You may have a strange result if you carelessly make an "EE" entry without first entering a 1. Try an operation to see how your calculator handles exponential entries that are not preceeded by a 1. The Texas Instruments TI59 gives the following product

2 "EE" 2 x "EE" 2 = 2 EE 22.

If the intent was to multiply $2 \times 10^2 \times 10^2$, the answer is rather far off.

Answers to Practice Problems

1. 4
2. 2
3. 5
4. 3
5. -4
6. -8
7. -7
8. -3
9. 0.00472
10. 0.000058
11. 0.00192
12. 0.0000072
13. 12×10^7
14. 36×10^{11}
15. 21×10^{21}
16. 6×10^{10}
17. 15×10^{-5}
18. 27×10^{-13}
19. 24×10^{-6}
20. 8×10^{-8}
21. 24×10^4
22. 18×10^{-6}
23. 24
24. 35×10^4
25. 0.5×10^2
26. 0.5×10^{10}
27. 0.2×10^6
28. 4×10^{12}

29. 2×10^{-6}

30. 3×10^{6}

31. 6×10^{12}

32. 3×10^{3}

33. 1.94×10^{5}

34. 101.2×10^{8}

35. 1.6×10^{-5}

36. 6.2×10^{-3}

37. 1.8543×10^{-3}

38. 16×10^{5}

39. 7.9×10^{5}

40. 6.4×10^{-17}

Exercises

1. $200{,}000 = 2 \times 10^{?}$
2. $13{,}576 = 13.576 \times 10^{?}$
3. $2387 = 238.7 \times 10^{?}$
4. $135 = 135{,}000 \times 10^{?}$
5. $12.6 = 126 \times 10^{?}$
6. $18.567 = 1.8567 \times 10^{?}$
7. $0.002 = 2 \times 10^{?}$
8. $0.00028 = 2.8 \times 10^{?}$
9. $0.0000387 = 38.7 \times 10^{?}$
10. $0.0000002158 = 0.2158 \times 10^{?}$

Express the following as decimals:

11. 2×10^{-3} = ______________
12. 2.3×10^{5} = ______________
13. 12×10^{7} = ______________
14. 12.6×10^{3} = ______________
15. 0.523×10^{4} = ______________
16. 0.0713×10^{6} = ______________
17. 16.23×10^{16} = ______________

18. 1.24×10^{-4} = ________________

19. 12.7×10^{-2} = ________________

20. 153.6×10^{-6} = ________________

21. 0.012×10^{-2} = ________________

22. 0.03×10^{-4} = ________________

23. 0.00004×10^{-6} = ______________

24. 0.123×10^{-5} = ________________

25. $(2 \times 10^{3}) \times (5 \times 10^{4})$ = ________

26. $(9 \times 10^{5}) \times (8 \times 10^{12})$ = ________

27. $(6 \times 10^{1}) \times (4 \times 10^{-9})$ = ________

28. $(3 \times 10^{-13}) \times (8 \times 10^{6})$ = ________

29. $(0.4 \times 10^{-9}) \times (0.2 \times 10^{-3})$ = ____

30. $(0.2 \times 10^{-7}) \times (0.2 \times 10^{12})$ = ____

31. $(3.2 \times 10^{-6}) \pm (4 \times 10^{-3})$ = _____

32. $(4 \times 10^{3}) \pm (8 \times 10^{5})$ = ________

33. $(6 \times 10^{5}) \pm (3 \times 10^{-9})$ = ________

34. $(3.2 \times 10^{-9}) \pm (8 \times 10^{-11})$ = ____

35. $(0.6 \times 10^{3}) \pm (0.2 \times 10^{6})$ = _____

Students often do not understand what to do with the quantities 10^1 and 10^0. 10^0 is one. This might occur by a multiplication such as $(2 \times 10^3) \times (3 \times 10^{-3}) = 6 \times 10^0 = 6$. The 10^0 is not arbitrary. $10^1 = 10$. For some applications 10^1 is preferred and for others 10 is preferred. Again, the definition is not arbitrary.

Before leaving this topic, it should be pointed out that the choice of scientific notation, engineering notation or some other powers of ten notation is really up to the user. It may appear to be cumbersome to write the number 38 as 3.8×10^1 or 0.38×10^2. However, there are people who choose to do exactly that. There are valid reasons for using scientific and engineering notation. You should know the rules and use the mode of expression best suited to the problem at hand. However, if your instructor has given you instructions on a specific notation, it is obviously to your advantage to follow those instructions.

3

Clearing Parentheses and Substitution of Numbers into Equations

The formulas of physics are a symbolic representation of phenomena that occur in nature. Eventually, for these representations to be useful, they must resolve to a numerical value for the quantities represented. Such numerical values are obtained by substitution of specific values of the variables represented by the symbols of a formula. The formula representing a phenomenon may be complex, and the substitution operation confusing. In this chapter, we will review the rules and procedures for substitution of numbers into formulas.

When substituting into formulas, it often helps to put the number substituted into parentheses or brackets. This is an extra step, but when signs are mixed, the extra step will reduce the chance of error.

The formulas of physics occur in many different forms. For the purposes of this chapter we will be concerned only with the basic operations of addition, subtraction, multiplication, and division.

3.1 ADDITION

Consider the equation

$$a + b = c.$$

This equation expresses in algebraic form the statement: "The sum of variable a and variable b is equal to the third variable c." Given the values of any two of the variables, the third may be found from the equation. Since we are concerned with addition, let us take $a = +2$ and $b = +3$. The substitution would be

$$(+2) + (+3) = c.$$

This would be stated: "Positive 2 plus positive 3 equals c." Note the value +2 is called "positive two" and not "plus two." The word plus implies addition, and the value of the variable being positive 2 means the variable is two more than zero on the variable scale.

When the parentheses are removed, (+2) becomes simply 2 (when no sign is written, it is understood that the value is positive) and (+3) becomes 3. The operation + (+3) does not change the sign, and (+3) = + 3.

Rule: When parentheses are removed following an addition operation, the sign of the quantity in the parentheses does not change.

The expression

$$(+2) + (+3) = c$$

becomes

$$2 + 3 = c, \text{ and}$$

$$c = 5.$$

As a little more complex example, consider the equation

$$a + (b + c^2) = d.$$

This equation has four variables, and if three are known the fourth may be determined. For the purposes of the example, take $a = 2$, $b = 3$, and $c = 5$ (note that the values are all positive, and the + is omitted for convenience). Substitution may be done in either of two ways: 1) Remove the parentheses and then make the substitutions. Or 2) make the substitutions and then remove the parentheses. In order to proceed, we will first remove the parentheses and make the substitutions.

$$a + (b + c^2) = d,$$

$$(2) + (3) + (5)^2 = d,$$

$$2 + 3 + 25 = d,$$

$$30 = d.$$

If substitution is done before the parentheses are removed:

$$(2) + ([3] + [5]^2) = d,$$

$$2 + (3 + 25) = d,$$

$$2 + (28) = d,$$

$2 + 28 = d,$

$30 = d.$

When substitution involves a negative quantity, the same procedure is followed. Going back to the original equation,

$a + b = c,$

let $a = 2$ (i.e., $+2$) and $b = -3$ (read as "negative three").

$(2) + (-3) = c.$

Applying the rule, this becomes

$2 - 3 = c$, and

$-1 = c.$

For $a = 3$, $b = -4$, and $c = -2$, the equation

$a + (b + c^2) = d$

becomes

$(3) + ([-4] + [-2]^2) = d,$

$3 + [-4] + [4] = d,$

$3 - 4 + 4 = d,$

$3 = d.$

3.2 SUBTRACTION

<u>Rule</u>: When a parentheses follows a minus sign, the sign of the quantity inside the parentheses changes. A positive quantity becomes negative and a negative quantity becomes positive.

The rule for a negative sign before parentheses is different from that for plus signs. To see how this works, consider the equation

$d - e = f,$

where $d = 2$ and $e = 3$ (both positive values). Substituting:

$d - e = f,$

$(2) - (3) = f.$

When the parentheses are removed, this becomes

$$2 - 3 = f,$$

$$-1 = f.$$

For the same equation, let $d = 2$ (positive) and $e = -3$.

Then

$$d - e = f,$$

$$(2) - (-3) = f.$$

The change of sign following the minus sign makes -3 become +3, and

$$2 + 3 = f,$$

$$5 = f.$$

The equation below is more complicated. Substitute the values $x = 3$, $y = -2$, $z = -3$, $d = -2$, and $e = -4$ into this equation.

$$x - (y - z) - (d + e) = f.$$

Although the equation is more complex, the procedures are exactly the same as before. Substitute the value of each variable, using parentheses or brackets to keep things straight:

$$(3) - ([-2] - [-3]) - ([-2] + [-4]) = f.$$

Remove the brackets,

$$(3) - (-2 +3) - (-2 -4) = f,$$

and simplify the terms in parentheses,

$$(3) - (1) - (-6) = f.$$

Finally, remove the parentheses and collect terms.

$$3 - 1 + 6 = f,$$

$$8 = f.$$

Of course, this could have also been solved by first removing the parentheses:

$$x - (y - z) - (d + e) = f,$$

$$x - y + z - d - e = f,$$

$$(3) - (-2) + (-3) - (-2) - (-4) = f,$$

$3 + 2 - 3 + 2 + 4 = f,$

$8 = f.$

Practice Problems

Solve for q in the following equations.

1. $q = +2 - (-3)$

2. $q = -3 + (+2 - 4)$

3. $q = (2 - 3) - (-8 + 4)$

4. $q = 4 + (2 - 3) - (-6 -8)$

5. $q = (2 - [3 - 4] + [-3 + (-2 + 1)]) +2$

6. $q = a - b + (-b)$, where $a = -2$ and $b = -1$

7. $q = x + (y - z) - (r - s)$, where $x = 2$, $y = 3$, $z = 4$, $r = -2$, and $s = -3$

8. $q = v - a - (-c + [s - t])$, where $v = -3$, $a = -5$, $c = 2$, $s = 1$, and $t = -3$

3.3 MULTIPLICATION

The notation of multiplication can be confusing. Quite often the symbol x is used as the multiplication operator. However, this symbol is also frequently used for a variable. (In this workbook, the symbol x will always represent a variable, and multiplication will be indicated by some other notation.)

There are several ways to indicate multiplication. If two variables represented by symbols are to be multiplied, they are most often written together. Thus, if a and b are to be multiplied, they are written as ab. Another multiplication instruction is to use adjacent parentheses. If the quantities $(a + b)$ and $(c + d)$ are to be multiplied, they are written $(a + b)(c + d)$.

When multiplying numbers, two methods are used to indicate multiplication. One is to enclose the numbers in parentheses, as in (3)(4); another is to put a dot between the numbers, as in 3·4. Of course, when the numbers are signed, parentheses are used, because the dot would be confusing; for example, (2)(-3).

Multiplication by signed numbers may or may not change the sign of the product, according to the following rule:

Rule: When two numbers of like sign are multiplied, the product is positive. When two numbers of unlike sign are multiplied, the product is negative.

To see how the rule works, consider the product of the numbers -3 and +4:

$$(-3)(+4) = -12.$$

Since the multiplication involves a product of two numbers of opposite sign, the product is negative. Similarly:

$$(+3)(-4) = -12.$$

The rule applies regardless of the order of the multiplication. Other examples are

$$(-3)(-4) = +12 \text{ (or simply 12)}.$$

$$(+3)(+4) = 12.$$

The same rules of multiplication apply when working with symbols. The only confusion arises when there is some question as to whether or not the numerical value of the symbol will ultimately be positive or negative. To keep things simple, all symbols are treated as positive. If the number substituted is negative, the equation will properly adjust to that fact according to the rules given earlier.

Rule: Symbols are always treated as positive quantities in equations. Positive or negative signs are assigned only when numerical values are substituted for the symbols.

Suppose you are working with an equation such as the following:

$$2\mu_0 E = \lambda.$$

Some of the symbols may not be familiar. In terms of handling the equation it makes no difference what the symbols are. These happen to be characters from the Greek alphabet that are often used in physics courses. The Greek letter μ is "mu" and λ is the letter "lambda." It may be disconcerting to see an unknown symbol, but whether or not the symbol is familiar has nothing to do with the manner in which the equation is handled.

Most physics equations involve more than one operation, mixing multiplication, addition, subtraction, and division. Again, a step-by-step approach will help avoid confusion. Suppose you have an expression such as the one below:

$$x[a(b - c) - (d - e)] = q.$$

Perhaps the most straightforward procedure is to start inside the brackets and work toward the outside. This will mean first remove the parentheses, then the brackets. Two operations involving parentheses are involved inside the brackets, one a multiplication and the other a subtraction.

When an expression enclosed by parentheses is multiplied by a quantity, each term of the parenthesized expression must be multiplied by the quantity.

Rule: Multiplication of an expression in parentheses by any quantity requires the multiplication of each term of the parenthesized expression by the quantity.

In the equation above, the product $a(b - c)$ is $ab - ac$. The subtraction of $(d - e)$ changes the sign of both d and e. The first simplification gives:

$$x[ab - ac - d + e] = q.$$

The final step is multiplication by x:

$$xab - xac - xd + xe = q.$$

If numbers are to be substituted, substitution can be done after the parentheses and brackets are cleared (as was done above) or before. Suppose $x = 2$, $a = 3$, $b = -4$, $c = 5$, $d = -2$, and $e = -3$. Direct substitution into the original equation gives

$$x[a(b - c)-(d - e)] = q,$$

$$(2)[(3)(\{-4\} - \{5\}) - (\{-2\} - \{-3\})] = q,$$

$$2[3(-4 - 5) - (-2 + 3)] = q,$$

$$2[3(-9) - (1)] = q,$$

$$2[-27 - 1] = q,$$

$$2[-28] = q,$$

$$-56 = q.$$

Substituting into the expression obtained by clearing the parentheses and brackets first gives:

$$xab - xac - xd + xe = q,$$

$$(2)(3)(-4) - (2)(3)(5) - (2)(-2) + (2)(-3) = q,$$

$$-24 - 30 + 4 - 6 = q,$$

$$-56 = q.$$

Multiplication of expressions in parentheses follows the preceding rule

$$(a + b)(c + d) = (a + b)c + (a + b)d = ac + bc + ad + bd.$$

Practice Problems

Solve for q in the following equations:

9. $q = -(2 - 3) + 4(2 -[-2])$

10. $q = (-2)[4 - 6(3 - 2)]$

11. $q = -6(2 + [-3]) + 2(3[-4 - 2(4 - 6) -2] + 10)$

12. $q = a + b[-c + d(e - f)]$, where $a = 1$, $b = -2$, $c = -3$, $d = -4$, $e = -1$, and $f = 4$

13. $q = (a + b)(c + d)$, where $a = -2$, $b = -1$, $c = 1$, and $d = -3$

14. $q = (a + b)x(a - b)$, where $a = -4$, $b = 2$, and $x = -5$

3.4 WORD PROBLEMS

A knowledge of the rules of mathematics and how to apply those rules is only part of what is involved in working physics problems. Hardly ever will you be given an equation to solve in a physics course. To the contrary, you will be given a statement about some physical situation and you must interpret what you are given and express the situation in the form of an equation. Only then can the tools of mathematics be brought to bear. If you cannot generate an equation from the statement of the problem, you cannot hope to find a solution.

At this state of review, the word problems that can be discussed are very simple. However, the procedures that are followed for even simple problems are the same as for the more complex. If you find word problems difficult, start with the simple steps below.

1. Read the problem until you are sure what is being asked. Don't start out until you know the direction you are going.

2. If necessary, make up a set of reasonable symbols to represent the variables given. The variables used are completely arbitrary, so choose symbols that are comfortable for you.

3. Use the information given to write a set of equations.

4. Solve the equations for the unknowns.

5. Examine your answer to see if it makes sense. If your answer is inconsistent with reality (for example, an object falls up, or a force is beyond reason), go back and examine your equations and the solution.

The following examples illustrate one mode of attacking a word problem.

Example 1: A van carries three kinds of boxes. Boxes of type A weigh 50 pounds, of type B 75 pounds, and of type C 100 pounds. The driver starts out the day with five boxes of type A, seven boxes of type B, and two boxes of type C. At his first stop, he delivers one box of each type and picks up two boxes of type B. At his second stop, he drops off three boxes, two of type A and one of type C. What is the weight of the boxes remaining in the van after the third stop?

Solution: One obvious nomenclature for the boxes is to let A represent the number of type A boxes, B the number of type B boxes, and C the number of type C boxes. The initial load (in pounds) in the van is

$$5A + 7B + 2C.$$

At the first stop the transaction is

$$-(A + B + C) + 2B.$$

At the second stop the transaction is

$$-2A - C.$$

The total boxes in the van is the number that the driver started with plus all of the transactions:

$$(5A + 7B + 2C) + (-[A + B + C] + 2B) + (-2A - C),$$

$$5A + 7B + 2C + (-A - B - C + 2B) - 2A - C,$$

$$5A + 7B + 2C - A - B - C + 2B - 2A - C.$$

Grouping all the A, B, and C together,

$$(5A - A - 2A) + (7B - B + 2B) + (2C - C - C),$$

$$2A + 8B.$$

The weight is 2(50) + 8(75) pounds, or 700 pounds.

Example 2: Fred makes \$8.50 per hour (exactly). He has \$35.00 (exactly). How much will he have if he works 15 hours?

Solution: Let x represent the number of hours that Fred works. He will earn (8.50)x dollars. His total money will be the money earned plus what he had to begin with, (\$8.50)x + \$35.00.

Substituting:

$$(\$8.50)(15) + \$35.00 = \$127.50 + \$35.00 = \$162.50.$$

We now have a question of significant figures. The least significant figure is the number of hours worked, and the question is whether Fred worked approximately 15 hours or exactly 15 hours. He could have worked as little as 14.5 hours or as much as 15.4 hours. He would correspondingly have $158.25 or $165.90. Unless given information to the contrary, we have to assume that the money Fred has is significant to only two digits, and the final correct answer is $160.00.

Methods of attacking more complex word problems are discussed in Chapters 5, 8, and 12.

Practice Problems

15. On a motoring trip, the Jones family drives 330 miles per day. How far from home are they after 5 days?

16. Sue is driving to the beach. At exactly 30 minutes after leaving home she is 25 miles from home. If she drives a steady 55 miles per hour and the beach is 118 miles from her home, how much time will be required to reach the beach from her home?

17. Joe works for $7.23 per hour. Before receiving his last check for 174 hours, he had $46.21. He owes taxes of 25% on all of his money. How much tax does he owe? (Taxing agencies tend to stretch the concept of significant figures when in their favor.)

Answers to Practice Problems

1. 5
2. -5
3. 3
4. 17
5. 1
6. 0
7. 0
8. 0
9. 17
10. 4
11. 14
12. -45
13. 6
14. -60
15. 1650 miles
16. 2.6 hours
17. $326

Exercises

Solve for x in the following equations.

1. $x = -(-3) + (+2)$

2. $x = 4 + (-2 + 1) - (3 - 4)$

3. $x = -(8 - 6) + (2 - 1)$

4. $x = -(-8) + (2 + 1) - (5 - 9)$

5. $x = a^2 - b + c$, where $a = -2$, $b = -5$, and $c = 8$.

6. $x = (-a)^2 + q - (y - z)$, where $a = -3$, $q = 5$, $y = -1$, and $z = -2$.

7. $x = z - (c - d) + (c + d)$, where $z = 7$, $c = 3$, and $d = -1$.

Solve for t in the following equations.

8. $t = -(-2 + 3) + (4 - 6[4 - 1])$

9. $t = (3 - [-5]) - 2([-2] + 1)$

10. $t = A + D (C - E[A + D])$, where $A = -2$, $D = 3$, $C = -5$, and $E = 4$.

11. $t = x - y(x - z)$, where $x = 3$, $y = -4$, and $z = -2$.

12. $t = q - 3\Omega (a - d)$, where $q = -5$, $\Omega = 4$, $a = 2$, and $d = -1$.

13. Joe has $2.35. How many canned drinks can he buy at 40 cents a drink, and how much money will he have left if he buys all that he can?

14. Sally runs a mile in 6.25 minutes. How long will it take Sally to complete a 10 mile run at this pace?

15. Fred's auto goes 25 miles on 1 gallon of gasoline. How much gasoline will be required if he makes a 220 mile trip?

4
Fractions

A fraction is a number resulting from a division. The unknown quantity x divided by y may be written as the term $\frac{x}{y}$ or x/y. The top number is called the numerator and the bottom number is called the denominator. Fractions may be used in any operation where any other number may be used.

4.1 REDUCING FRACTIONS TO LOWEST TERMS

When the numerator and denominator of a fraction contain a common factor, the fraction may be reduced (simplified) by removing that factor. The value of a fraction is not changed so long as the top and bottom are divided by the same factor.

> Rule: The value of a fraction is not changed so long as the numerator and denominator of the fraction are multiplied or divided by the same factor.

The fraction $\frac{9}{12}$ contains the common factor 3 in both numerator and denominator. If the fraction is divided top and bottom by 3, the fraction is reduced to $\frac{3}{4}$, or 3/4:

$$\frac{3\cdot 3}{3\cdot 4} = \frac{3}{4}$$

The process of dividing top and bottom by a common factor is often called cancelling. For example, when the fraction $\frac{9}{12}$ is written as $\frac{3\cdot 3}{3\cdot 4}$, the process of cancelling is indicated with a slash:

$$\frac{\not{3}\cdot 3}{\not{3}\cdot 4}.$$

Cancelling must be done correctly. The cancelling process in invalid unless every term in the numerator and denominator contains the factor to be cancelled. The fraction $\frac{3x + 12}{4}$ cannot be reduced to $3x + 3$. This very common incorrect result is obtained by dividing the 12 by 4 without dividing the 3x by 4.

$$\text{WRONG:} \quad \frac{3x + 12}{4} = \frac{3x + 3\cdot\not{4}}{\not{4}} = 3x + 3.$$

It may be that you wish to divide out the denominator. In the case of $\frac{3x + 12}{4}$, the correct final result would be $(\frac{3x}{4}) + 3$.

Other examples of common incorrect cancelling operations are

$$\frac{4x + 5}{2} = 2x + 5$$

$$\frac{3}{6x - 7} = 1/(3x - 7).$$

Rule: Factors may be cancelled only when numerator and denominator are products. Terms appearing as sums or differences may not be cancelled unless each term in the sum or difference contains a common factor to both the numerator and denominator.

Under this rule, the fraction $\frac{ax + ab}{2ay - ab}$ may be factored to the fraction $\frac{a(x + b)}{a(2y - b)}$, and reduced to $\frac{x + b}{2y - b}$ by cancelling the a.

$$\frac{\not{a}(x + b)}{\not{a}(2y - b)} = \frac{x + b}{2y - b} .$$

No further reduction is possible. The b terms do not cancel.

Fractions may be written with a slash (/) as well as numerator over denominator. This often happens in materials written for handouts or on examinations because it is easier to type 3/4 than $\frac{3}{4}$. The fraction is handled exactly the same way, whether written with the / (as x/y) or as numerator over denominator (as $\frac{x}{y}$).

Practice Problems

Reduce the following fractions to lowest terms.

1. $\frac{6}{18}$

2. $\frac{49}{56}$

3. 108/126

4. $\frac{3a^2b}{6a^3b^2y}$

5. $\frac{21xy^3}{3xy}$

6. $28pr^2s/14apx$

7. $\frac{2x + 18}{6}$

8. $\dfrac{3z - 4}{z}$

9. $(ax + ay)/(bx + by)$

4.2 NEGATIVE FRACTIONS

A negative sign in a fraction may be attached to the numerator, the denominator or to the fraction as a whole:

$$\frac{-a}{b} = -\frac{a}{b} = \frac{a}{-b}.$$

When two negative signs are present, the resulting sign is +:

$$-\frac{-a}{b} = \frac{a}{b}; \quad \frac{-a}{-b} = \frac{a}{b}; \quad -\frac{a}{-b} = \frac{a}{b}.$$

If three negative signs are present, the resulting fraction is negative:

$$-\frac{-a}{-b} = \frac{-a}{b}.$$

The most general convention is to write negative fractions with a negative numerator. However, there is no fundamental reason for this convention, and the negative sign may equally well be assigned to the denominator or the fraction as a whole.

4.3 MULTIPLICATION OF FRACTIONS

As the symbol x is often used as a variable, multiplication is best indicated by some convention other than the use of this symbol. One way of doing this is to use parentheses, $(\frac{a}{b})(\frac{c}{d})$, or a dot, such as $\frac{a}{b} \cdot \frac{c}{d}$.

Regardless of notation, when fractions are multiplied the product is a fraction whose numerator and denominator are the respective products of the numerators and denominators of the original fractions.

Rule: Fractions are multiplied by multiplying numerators to produce the numerator of the product fraction and multiplying denominators to produce the denominator of the product fraction.

$$\frac{x}{y} \cdot \frac{a}{b} = \frac{ax}{yb}$$

After multiplication, the product should be reduced to lowest factors.

For example, the product of $\frac{9}{28}$ and $\frac{14}{6}$ is

$$\frac{9}{28}\cdot\frac{14}{6} = \frac{126}{168} = \frac{3\cdot\cancel{3}\cdot\cancel{2}\cdot\cancel{7}}{2\cdot2\cdot\cancel{7}\cdot\cancel{2}\cdot\cancel{3}}$$, which reduces to $\frac{3}{4}$.

Practice Problems

Multiply the following fractions and reduce to lowest factors.

10. $\frac{2}{3}\cdot\frac{5}{7} =$ ____________________

11. $\frac{2}{9}\cdot\frac{27}{8} =$ ____________________

12. $(3/25)(1/21) =$ ____________

13. $\frac{4b}{3xy}\cdot\frac{3y}{2ab} =$ ____________

14. $(2x^2z/3b)\cdot(6a^2b/xyz) =$ _____

15. $(x^2za/2y^2)(y/2ax) =$ ________

4.4 DIVISION BY FRACTIONS

When one fraction is divided by another, the divisor is inverted and multiplied.

Rule: To divide by a fraction, invert the fraction and multiply. $\frac{a}{b}\div\frac{x}{y} = \frac{a}{b}\cdot\frac{y}{x} = \frac{ay}{bx}$.

Practice Problems

Divide the following fractions and reduce to lowest factors.

16. $\frac{2}{3}\div\frac{1}{2} =$ ____________________

17. $\frac{3}{14} \div \frac{1}{4} =$ ________________

18. $(12/6) \div (2) =$ ___________

19. $(x^2y/z) \div (x/y) =$ ________

20. $\frac{3a^2b}{2c} \div \frac{ac}{ab} =$ ______________

21. $\frac{x^2 + 2x}{x + 4} \div \frac{x + 2}{x - 4} =$ ________

4.5 ADDING AND SUBTRACTING FRACTIONS

When fractions are added, the numerator and denominator are not treated the same. Fractions cannot be added unless they have the same denominator, and in the addition process the denominator does not change. The addition operation produces a fraction whose numerator is the same as the sum of the original fractions and whose denominator is the same as the original fractions.

Rule: Fractions with the same denominator add by adding the numerators and retaining the denominator according to

$\frac{x}{A} + \frac{y}{A} = \frac{x + y}{A}$. Fractions cannot be added unless they have the same denominator.

The correctness of the rule may be seen by adding two simple fractions such as $\frac{1}{4}$ and $\frac{3}{4}$. One quarter plus three quarters is one, and this is found by the addition:

$$\frac{1}{4} + \frac{3}{4} = \frac{1 + 3}{4} = \frac{4}{4} = 1.$$

If both numerators and denominators are added, the result is $\frac{1 + 3}{4 + 4}$, or $\frac{4}{8} = \frac{1}{2}$, which is not correct.

Subtraction of fractions is accomplished by adding the negative of the fraction. Since the addition takes place in the numerator, it is most convenient to put the negative sign with the numerator.

$$\frac{1}{2} - \frac{3}{2} = \frac{1}{2} - \frac{3}{2} = \frac{(1 - 3)}{2} = \frac{-2}{2} = -1.$$

Practice Problems

22. $\frac{1}{4} + \frac{1}{4} =$ ____________________

23. $\frac{4}{3} + \frac{7}{3} =$ ____________________

24. $\frac{1}{5} - \frac{6}{5} =$ ____________________

25. $\frac{4}{7} - \frac{3}{7} =$ ____________________

4.6 EQUIVALENT FRACTIONS AND COMMON DENOMINATORS

If the fractions to be added or subtracted do not have the same denominator, they must be converted to equivalent fractions. Multiplying the numerator and denominator of a fraction by the same number does not change the value of the fraction. If we multiply numerator and denominator of each fraction by the appropriate factor, after the multiplication the fractions will have the same denominator. To see how this works, consider

$\frac{1}{2}$ and $\frac{1}{3}$. If $\frac{1}{2}$ is multiplied by $\frac{3}{3}$ and $\frac{1}{3}$ by $\frac{2}{2}$, the new fractions will be

$\frac{3}{6}$ and $\frac{2}{6}$ respectively. The fractions now have the common denominator 6.

$$\frac{1}{2} + \frac{1}{3} = \frac{1}{2} \cdot \frac{3}{3} + \frac{1}{3} \cdot \frac{2}{2} = \frac{3}{6} + \frac{2}{6} = \frac{(3 + 2)}{6} = \frac{5}{6}$$

Practice Problems

26. $\frac{1}{2} + \frac{1}{5} =$ ____________________

27. $\frac{2}{5} - \frac{1}{7} =$ ____________________

28. (3/2) + (2/5) = ____________

29. (11/3) + (5/11) = __________

30. $\frac{3}{x} - \frac{2}{y} =$ ____________________

31. $\frac{a}{b} + \frac{c}{d} =$ ____________________

When the denominators of the fractions to be added contain common factors, it is most efficient to convert each to an equivalent fraction with the lowest common denominator. The <u>lowest common denominator</u> is the smallest number that contains all of the factors in the denominators of the fractions being added. Of course, if the denominators do not contain any common factors, the lowest common denominator is the product of the original denominators, as with the examples and practice problems to this point.

The lowest common denominator must contain every factor in the denominators as many times as it appears in any of the denominators. For example, $\frac{1}{3}$ and $\frac{1}{12}$ have denominators with factors (3) and (2•2•3) respectively. The lowest common denominator must contain both 3 and 2•2. The lowest common denominator will be 12.

$$\frac{1}{3} + \frac{1}{12} = \frac{1}{3} \cdot \frac{4}{4} + \frac{1}{12} = \frac{4}{12} + \frac{1}{12} = \frac{5}{12}.$$

For more complex denominators, the procedure is the same as with the simple example above. Consider the fractions $\frac{2x}{y^2z}$ and $\frac{3}{4xyz^2}$.

To add these fractions, the first step is to find the lowest common denominator. Each denominator is written with all its factors, and the lowest common denominator is found by using each factor the maximum number of times that it occurs in either denominator.

$$y^2z = y \cdot y \cdot z$$

$$4xyz^2 = 2 \cdot 2 \cdot x \cdot y \cdot z \cdot z.$$

The y appears twice in the first and once in the second, so y^2 is one factor for the lowest common denominator. z appears once in the first and twice in the second, so z^2 is in the lowest common denominator. 4 (as 2·2) is not in the first but is in the second, so 4 is in the lowest common denominator. Finally, x does not appear in the first but does in the second, so x is also a factor in the lowest common denominator. Putting all of these together, the lowest common denominator will be y^2z^24x, or $4xy^2z^2$. To add the original fractions:

$$\frac{2x}{y^2z} + \frac{3}{4xyz^2} = \frac{2x}{y^2z} \cdot \frac{4xz}{4xz} + \frac{3}{4xyz^2} \cdot \frac{y}{y} = \frac{8x^2z}{4xy^2z^2} + \frac{3y}{4xy^2z^2} =$$

$$\frac{8x^2z + 3y}{4xy^2z^2}$$

Stated in this form, the fraction is in lowest terms and no further simplication is possible.

Practice Problems

Assume the expressions below are denominators and find the lowest common denominator.

32. 3 and 5

33. 3 and 12

34. 6 and 24

35. 6 and 27

36. 12 and 33

37. 7 and 15

38. y^2 and 2y and $(y + 2)^2$

39. a^2 and ab and ab^2

Add or subtract the fractions as indicated.

40. $\frac{1}{5} + \frac{1}{15} =$ ______________

41. $\frac{1}{14} + \frac{1}{35} =$ ______________

42. $\frac{1}{A} + \frac{1}{B} =$ ____________________

43. $\frac{2x + 12}{6} + \frac{3}{x} =$ ______________

44. $(x - 1)/(x + 1) - (y + 1)/(y - 1) =$ ____________________

45. $a^2/(a + b) - b^2/(a - b) =$ ______________________________

Answers to Practice Problems

1. 1/3
2. 7/8
3. 6/7
4. 1/2aby
5. $7y^2$
6. $2r^2s/ax$
7. (x + 9)/3
8. (3z - 4)/z
9. a/b
10. 10/21
11. 3/4
12. 1/175
13. 2/ax
14. $4a^2x/y$
15. xz/4y
16. 4/3
17. 6/7
18. 1
19. xy^2/z
20. $3a^2b^2/2c^2$
21. x(x - 4)/(x + 4)
22. 1/2
23. 11/3
24. -1
25. 1/7
26. 7/10
27. 9/35
28. 19/10
29. 136/33
30. (3y - 2x)/xy
31. (ad + cb)/db
32. 15
33. 12
34. 24
35. 54
36. 132
37. 105
38. $2y^2(y + 2)^2$
39. a^2b^2
40. 4/15
41. 7/70 = 1/10
42. (A + B)/AB

43. $(x^2 + 6x + 9)/3x$

44. $-2(x + y)/(xy + y - x - 1)$

45. $(a^3 - a^2b - b^2a - b^3)/(a^2 - b^2)$

Exercises

Reduce to lowest terms:

1. $21/77$
2. $35/63$
3. $96/156$
4. $144/200$
5. $6ab^2/21a^2c$
6. $14p^2\Omega/7q^3v^2x$
7. $(a^2 - b^2)/(2a + 2b)$
8. $(a^2x + a^2)/(z + xz)$

Multiply and reduce to lowest terms:

9. $\frac{15}{21} \cdot \frac{35}{25}$
10. $\frac{6a^2b}{13b^2c} \cdot \frac{5c^3}{12a^3b^2}$
11. $(5a^2b^3/3dc)(21a^3bc/10d)$
12. $(x^2y/3z^3)(20z^2x/4qy^2)$
13. $[(ab^2 - ac)/d^2][cd/(ac - ax^2)]$
14. $[(x^2 - y^2)/xz][(x - y)/(zx + zy)]$

Divide and reduce to lowest terms:

15. $\frac{3}{5} \div \frac{25}{27}$
16. $(49/16) \div (35/28)$
17. $(ab^2x/yz) \div (ax^2y/b^2x)$

18. $[(2a^2 + a)/(2b^2 - 8)] \div [(6a + 3)/(4b + 8)]$

Perform the indicated operation and reduce to lowest terms

19. $\frac{2}{7} + \frac{3}{25} =$ ____________________

20. $\frac{5}{29} + \frac{1}{2} =$ ____________________

21. $\frac{2}{12} - \frac{1}{48} =$ ____________________

22. $\frac{7}{30} - \frac{11}{45} =$ ____________________

23. $1/(a - b) - 1/(a + b) =$ ____

24. $\frac{x + 4}{2x} + \frac{2x}{x + 4} =$ ____________

5

Equations with One Unknown

Physics utilizes mathematical expression because it would be too cumbersome to do it any other way. As you progress in the physics course, you will come to "read" an equation (often called a formula) much in the same way that you read a verbal sentence. In fact, the process of converting a physics statement to a mathematical equality is often called translation. This translation is just as real as translation of one spoken or written language to another, and, just as with spoken or written language, fluency comes with practice.

Beginning students often look upon physics as just one equation to be memorized after another. While it is true that equations are very important, always keep in mind that these equations represent descriptions of physical events. Languages of any type are used to communicate ideas. The formulas of physics are a convenient mechanism to express the ideas and concepts.

Of course, this does not degrade the importance of being able to use the mathematics. To the contrary, until you are proficient to the point of carrying out the mathematical operations with ease, you cannot look beyond the mathematics to what the equations represent physically. In this chapter equations containing one unknown are presented. In the physics course, similar equations might represent the location of something at a given time under a specified set of circumstances, or perhaps state the time at which something specific will happen. For any set of conditions where everything is set except for one unknown, the methods of this chapter will be required.

5.1 LINEAR EQUATIONS

An equation containing two variables related in a manner such that a plot of one against the other results in a straight line is called a linear equation. The relationship between the variables is correspondingly called a linear relation.

Linear relations are very common in physics. For example, in the study of motion, if the velocity is constant, the distance traveled is linear in time. If you travel at a constant speed of 30 miles per hour, you will go 30 miles in one hour, 60 miles in two hours, 90 miles in three hours, etc. A plot of distance versus time will be a straight line. If the acceleration is constant, distance traveled will no longer be linear in time, but velocity will. As the algebra of motion is usually one of the first topics studied in a physics course, you very likely will encounter linear equations in the first week or so of your physics class.

A linear equation has the form

$$y = mx + b,$$

where m and b are constants and x and y are the variables. Graphs are discussed in detail in Chapter 10, but for the time being accept the use of the words "slope" and "intercept" as constants. The constant m is the slope of the line when x is plotted on a horizontal axis against y on a vertical axis. The variable b is given the name intercept and is the value of the variable y when the variable x is equal to zero.

Often, when one variable is written as equal to another in a functional form, it is said to be "dependent" on the second variable. In the equation above, since it is written "y = something having to do with x and a constant," y is called the dependent variable. The second variable, x in the equation above, is called the independent variable.

However, in any equation involving two variables, either one can be turned into the "dependent" variable by manipulation of the equation. To physics, the designation of one as independent and the other as dependent has little meaning. If you are studying motion, you might use a certain equation in one problem to locate where something is at a particular time. In the next problem, you may very well use the same equation to determine when an object will be at a particular place. The formulas are presented in what has become a traditional format, where one variable is equal to some function of the other. However, this is not restrictive, making one variable always be dependent on the other. The relation between the variables is given by a certain equation. Your task will be to find either one, given the other.

For example, suppose that in the equation

$$y = mx + b,$$

$y = 4$, $m = 8$, and $b = -12$. The unknown is x, not y. One method of approach is to find the equation for x in terms of y, making x the dependent variable and y the independent variable, and then substituting the numerical values given to solve for x. With much less work, you can solve directly for x.

Making the substitutions,

$(4) = (8)x + (-12).$

Clearing the parentheses gives

$4 = 8x - 12.$

The unknown is x, and the goal is to reduce this equation to an expression of the form x = a number. To accomplish this, we must rearrange the terms in the equation without destroying the equality. This is done according to the following rule.

Rule: Equality is maintained so long as the same thing is done to both sides of an equation.

We have

$4 = 8x - 12$

and wish to solve for x. If we add 12 to both sides of the equation, we will be left with x and its multiplier on one side only.

$(12) + 4 = 8x - 12 + (12),$

$16 = 8x.$

Next, we divide both sides by 8 to leave x standing alone on one side of the equal sign.

$16/(8) = 8x/(8),$

$2 = x.$

For the equation $y = mx + b$, when $y = 4$, $m = 8$, and $b = -12$, x will be equal to 2.

Frequently you will have to solve equations treating symbols as if they are constants. Sometimes students find this disturbing, with the result that problems requiring manipulation of symbols are considerably more difficult than when only numbers are involved. Even students with substantial training in mathematics may go into "symbol shock" when they have to manipulate symbols. You will learn to work with symbols in physics. The "formulas" of physics are equations which describe the relation between two (or more) variables under different conditions, where the different conditions as well as the variables are represented by symbols. As the formulas apply to a variety of conditions, so the conditions must be represented by symbols. A student of physics must be able to handle the formulas well, without regard to whether they contain numbers or symbols.

Symbol shock is amplified by the use of unfamiliar characters. Greek letters are often used in physics tests. For example, in the study of rotational motion you will likely encounter the equation

$$\omega = \omega_0 + \alpha t$$

(ω is the character "omega" and α is the character "alpha" from the Greek alphabet. The subscript o on ω_0 means that this is a special value of the variable ω).

If you are asked to solve this equation for α, you follow the same steps as with the numerical example given earlier. First isolate the terms containing α by subtracting ω_0, from both sides of the equation:

$$\omega - \omega_0 = \omega_0 + \alpha t - \omega_0,$$

$$\omega - \omega_0 = \alpha t.$$

Dividing both sides by t gives α:

$$\frac{\omega - \omega_0}{t} = \frac{\alpha t}{t} = \alpha.$$

If you wish, the equation may be written with the desired unknown α on the left:

$$\alpha = \frac{\omega - \omega_0}{t}.$$

Practice Problems

Solve for the variable in the equations below.

1. $2x = 4$
2. $2x = 6 - 2$
3. $2z - 2 = 6$
4. $3 - 4q = -5$
5. $2 = 3a + 1$

Solve for x in the following equations.

6. $2x = 4y - 4$
7. $6x - 4 = 3z + 2$
8. $a - b = 2x + 3$
9. $\alpha - \beta = \theta x + s$
10. $\mu + \lambda\Omega = \pi x + \Delta$

Often the variable for which you are solving is found on both sides of the equal sign. The rule still works.

For example, consider

$$4 - 2x = 6(x - 2).$$

The first step is obvious: Clear the parentheses.

$$4 - 2x = 6x - 12.$$

The next goal is to put all terms containing x as a factor on one side of the equation and everything else on the other. This is a two-step process, one for the terms containing x and one for the other terms. Either may be done first. In order to proceed, we will first add 2x to both sides:

$$4 - 2x + (2x) = 6x - 12 + (2x),$$

$$4 = 8x - 12.$$

Now the 12 can be removed from the left side:

$$4 + (12) = 8x - 12 + (12),$$

$$16 = 8x,$$

$$2 = x.$$

For equations with symbols, the procedure is exactly the same. Consider the equation

$$Ax + B = ax + b.$$

Subtract ax from each side:

$$Ax + B - (ax) = ax + b - (ax),$$

$$Ax - ax + B = b,$$

$$x(A - a) + B = b.$$

Subtract B from both sides:

$$x(A - a) + B - B = b - B,$$

$$x(A - a) = b - B.$$

Finally, divide by the multiplier of x:

$$x = \frac{b - B}{A - a}.$$

Practice Problems

Solve for the variable in the following equations.

11. $12x - 2 = 2x$

12. $2 - 2z = -6z + 3$

13. $3(2 - 5a) = 3 + 2a$

14. $3b + 7 = 2(b - 2)$

Solve for x in the following equations.

15. $x - y = 5x + 2$

16. $3x - 2y = 3z + 2x + 3$

17. $A - Bx = C + D(x + F)$

18. $\alpha - \beta x = \delta(x - \alpha)$

19. $\theta - \omega x = \alpha x - \omega$

5.2 EQUATIONS CONTAINING QUADRATIC TERMS

When an equation contains a variable to the second power (squared) as the highest power in the equation, the equation is called a quadratic equation in that variable. For example, consider an equation that occurs in the study of motion:

$$x = x_0 + v_0 t + (1/2)\ at^2.$$

This equation contains several impressive looking variables, all of which mean something to physics. More to the point, not all the quantities that could be variables are linear. Specifically, the equation is quadratic in t. In this chapter, we are concerned only with linear variables, so t will always be a given (or treated as a constant). Chapter 7 is devoted to the solution of quadratic equations for the squared variable.

Given the constraint that we are concerned only with linear variables, the form of the quadratic is more of a distraction than the introduction of anything new. Consider the equation

$$4 = 2 + st + 6t^2,$$

where $t = 2$.

Substituting:

$$4 = 2 + s(2) + 6(2)^2,$$

$$4 = 2 + 2s + 24.$$

The equation becomes one solving for the variable s in equations much like the ones discussed earlier. Simplifying:

$$4 = 26 + 2s.$$

Subtracting 26 from each side gives

$$4 - (26) = 26 + 2s - (26),$$

$$-22 = 2s.$$

Dividing both sides by 2, we have

$$-(22)/2 = (2s)/2,$$

$$-11 = s.$$

Most likely you eventually will have to work with symbols in the quadratic equations, and if so they are handled just as in linear equations. As an example, let us solve for a in the equation given at the start of this section:

$$x = x_0 + v_0 t + (1/2)\ at^2.$$

Subtract $(x_0 + v_0 t)$ from both sides.

$$x - (x_0 + v_0 t) = x_0 + v_0 t + (1/2)\ at^2 - (x_0 + v_0 t),$$

$$x - x_0 - v_0 t = x_0 + v_0 t + (1/2)\ at^2 - x_0 - v_0 t,$$

$$x - x_0 - v_0 t = (1/2)\ at^2.$$

Multiply by 2,

$$2(x - x_0 - v_0 t) = 2(1/2)\ at^2,$$

$$2(x - x_0 - v_0 t) = at^2,$$

and divide by t^2,

$$\frac{2(x - x_0 - v_0 t)}{t^2} = \frac{at^2}{t^2},$$

to finally obtain

$$\frac{2(x - x_0 - v_0 t)}{t^2} = a.$$

Practice Problems

Solve for the remaining variable in the equations below using the given value for t.

20. $t = 2$; $4 = a - 3t + (1/2)(6)t^2$

21. $t = 0.5$; $x = -4 + 6t + (1/2)(2)t^2$

22. $t = -3$; $3 = 5 - vt - (1/2)(7)t^2$

5.3 WORD PROBLEMS

Word problems require two separate and distinct operations. The first is to read the problem for information necessary to write an equation (or equations) and the second is to solve the equation(s) for the unknown(s). Since this chapter is devoted to equations with only one unknown, the word problems presented will have only one equation.

In working word problems, it is extremely important that you understand the problem before attempting to solve it. The odds of finding a solution by accident are very slim, and unless you are sure about what you are looking for and the direction your search is going to take, word problems will be a source of frustration. Although the problems of this chapter are simple (because the mathematics to this point in this work book has been simple), you should work through them if you are not comfortable with word problems. They will give you practice with putting information into an equation where you have generated the equation using your variable to represent the unknown.

As an example of how you attack a word problem, consider the following:

> A girl scout sells cookies at $2.00 per box. When the girl starts the sales, she has $12.00. How many boxes of cookies must the girl sell to have a total of $34.00?

The solution starts with an understanding that we are looking for how many boxes of cookies the girl must sell. Let the variable x represent that unknown. At $2.00 per box, when she sells x boxes she will have ($2.00)x. The money she then has will be what she started with plus what she has made from sales. If the total money she has is $34.00, the equation is

$$\$34.00 = \$12.00 + (\$2.00)x.$$

The equation is solved the same way as the other equations of this chapter.

Subtracting $12.00 from each side gives

$$\$22.00 = (\$2.00)x,$$

and dividing by $2.00 gives

$$x = 11.$$

When the girl sells 11 boxes, she will have \$22.00. This money added to the original \$12.00 results in her having \$34.00.

The final step in working a word problem (or a physics problem) is to examine the solution to see if it makes sense. This doesn't mean to go back over the equations. Rather it means to think about the solution to see if the answer is reasonable. In the scout cookies example, the girl will generate \$22.00 from sales. She already had about half that amount, so if 11 isn't the exact answer it will be close. If you had made a mistake in setting up the equation, and a number like 2 or 100 had resulted from your solution, your algebra may well have been correct, but since the equation would not have been the proper one, the solution to the equation would not have been the answer. Often when the equation is wrong, the solution is so far off that a quick examination of the answer to see if it is reasonable is adequate.

In physics problems, it is often possible to have an approximate idea of the answer before starting the problem. You have some experience with many of the parameters used in problems, so a few seconds spent thinking about the problem and getting an idea of the numerical value of the solution helps alert you to mathematical errors. In any event, always examine your solution for reasonableness.

Practice Problems

23. Your Yummie Baking company bakes a certain number of cakes every month except in December when they bake twice as many cakes as normal. If they baked 500 cakes last December, how many do they bake in a normal month?

24. Bill has a certain amount of money in his bank account. If he were to double his money, he would have \$250 less than \$1000. How much money does he have now?

Answers to Practice Problems

1. $x = 2$
2. $x = 2$
3. $z = 4$
4. $q = 2$
5. $a = 1/3$
6. $x = 2y - 2$
7. $x = z/2 + 1$
8. $x = (a - b - 3)/2$
9. $x = (\alpha - \beta - s)/\theta$
10. $x = (\mu + \lambda\Omega - \Delta)/\pi$
11. $x = 1/5$
12. $z = 1/4$
13. $a = 3/17$
14. $b = -11$
15. $x = -y/4 - 1/2$
16. $x = 2y + 3z + 3$
17. $x = (A - DF - C)/(B + D)$
18. $x = \alpha\ (1 + \delta)/(\beta + \delta)$
19. $x = (\theta + \omega)/(\alpha + \omega)$
20. $a = -2$
21. $x = -0.75$
22. $V = 59/6$
23. 250
24. \$375

Exercises

Solve for the variable in the equations below.

1. $3 = 2w - 4$

2. $b = 2b + 5$

3. $x = 24 - x$

4. $12x - 2 = 4$

5. $-6 -3z = 8$

6. $2 - 2p = -6$

7. $6 - 4q = 10$

Solve for x in the equations below.

8. $\alpha - \beta = 2x$

9. $\theta - s = \theta x + 3$

10. $x - y = 5x -2$

11. $3x - 2y = 3z + 2x$

12. $A + Bx = c + Dx$

13. $a - b(x - 4) = 2x - 5$

14. $\alpha - v_0xt = 2t(4 - 3x)$

15. $(2 - 5x)(a + b) = a(4x - b)$

16. $(2 - 5x)[2 - (-c)] = (5 + t)(x - 4)$

17. $[a - bx + (-x)](2b - 4) = (a - b)(b - x)$

18. $(A + xy)(Ax - y) = xy(2y + Ax)$

Solve for the unknown in the equations below using the given value for t.

19. $t = -2$; $q = -6 + 3t + (1/2)(4)t^2$

20. $t = 3$; $a = 3 - 4t - 6t^2$

21. $t = 0.5$ $3z = -4t - 5tz + 3zt^2$

22. $t = 2$ $4t^2 = -3r + 4rt - 2t^2$

Solve for q in the equations below.

23. $x = q + at + bt^2$

24. $s = r + qt - at^2/2$

25. $v = u - vt + qt^2/2$

26. $A = -B + qt - at^2$

27. $x = y - qt + xyt^2$

Answer the following word problems.

28. A car dealership sells each car for $500 more than it cost. The weekly expenses for the dealership are $5000. How many cars must be sold in a week to make a profit of $12,000?

29. Three consecutive integers sum to 27. What is the smallest number?

30. Sue bought a skirt and a blouse. The skirt cost twice as much as the blouse and the total for the two items was $12. What was the price of the skirt?

31. A father is 42 years old. His present age is three times what his daughter's age was 8 years ago. How old is the daughter?

32. Four times a number minus 6 is equal to two times the number. What is the number?

33. Five more than a certain number is equal to 17. What is the number?

34. Four plus 5 times a number is equal to 3 times 2 less than the number. What is the number?

35. Three consecutive numbers add to 12. What is the middle number?

36. An airplane flies at a speed of 300 miles per hour relative to the ground. The airplane starts out at a field 60 miles north of Bigtown to fly to Homesburg. Homesburg is 600 miles north of Bigtown. How long does it take for the airplane to reach Homesburg from the airfield?

37. When a certain number is tripled, the resultant is 3 less than 330. What is the original number?

38. The formula relating Celsius to Farenheit temperatures is $C = (5/9)(F - 32)$. What is the Celsius temperature when $F = 16°$?

6

Square Roots

Up to this point all of the problems in this workbook have involved only integers or ratios of integers. Even when working with symbols, the operations involved only multiplication, division, addition, and subtraction of these symbols exactly as if they were numbers.

Simple whole numbers and ratios of simple whole numbers are called rational numbers. Another class of numbers exists called irrational numbers. Irrational numbers cannot be expressed as ratios, products, sums, or differences of rational numbers.

For example, in the equation

$$x^2 = 6,$$

x is a number that when multiplied by itself will give the number 6. We call the value of x the square root of 6, and use the notation

$$x = \sqrt{6}.$$

x is a real number, and has the value

$$x = 2.449489743...,$$

where the integers to the right may be written until one tires of calculating (or more likely to the limit of the display of the calculator one uses). The irrational nature of the number is dictated by the fact that regardless of the accuracy of the value (that is, how many places) the pattern of numbers never repeats.

This is in contrast to the case for fractions. The value of π is often approximated by the fraction 22/7. Actually π is an irrational number and has value

$$\pi = 3.141592654\ldots,$$

and the fraction 22/7 is

$$22/7 = 3.12428571428571\ldots,$$

where the accuracy of the fraction is extended to illustrate why it is called a rational number. Notice that after the first three digits, the sequence 428571 repeats. This sequence will continue to repeat for as far as the calculation is taken. However, no sequence of digits for π will repeat, regardless of how far the sequence is carried.

The square root of 6 being an irrational number, it is impossible to write out an exact expression. Instead such a square root is indicated by using the radical, in this case $\sqrt{6}$. The value of the term $\sqrt{6}$ (which is 2.449489743...) when used in calculation, must be rounded to the appropriate number of places consistent with the correct number of significant figures.

The symbol $\sqrt{\ }$ is called the radical sign or simply the radical. As an extension of this nomenclature, a square root is often called a radical.

When a square root is taken, as in

$$x^2 = 6,$$

there are actually two numbers that satisfy this equation. Since multiplication by two quantities having the same sign gives a plus product, the two values of x that satisfy the equation are $x = +\sqrt{6}$ and $x = -\sqrt{6}$. (The + sign is usually implied.) Another way of writing this is $x = \pm\sqrt{6}$, read "x equals plus or minus the square root of six."

For any number x, the square root is defined by

$$\sqrt{x^2} = \pm x.$$

Radicals are handled in equations exactly as any other quantity. They may be left as radicals or approximated as numbers as the situation dictates.

6.1 SIMPLIFYING RADICAL EXPRESSIONS

Rule: The square root of a product is equal to the product of the square roots of the factors. $\sqrt{ab} = \sqrt{a}\,\sqrt{b}$

Complex radicals are simplified by the above rule.

$$\sqrt{a^2b} = \sqrt{a^2}\sqrt{b} = a\sqrt{b}.$$

While literally the above solution is $\pm a\sqrt{b}$, the usual practice is to leave the $\pm$ off and write the answer simply as $a\sqrt{b}$, as above.

Practice Problems

Simplify the following expressions.

1. $\sqrt{6ab^3}$ = ______________________

2. $\sqrt{8x4y^2}$ = ______________________

3. $\sqrt{2xy}\ \sqrt{xy^2}$ = ______________________

4. $\sqrt{3z}\ \sqrt{2q^3}\ \sqrt{6zq}$ = ______________________

5. $\sqrt{3xA}\ \sqrt{24y}\ \sqrt{AB}$ = ______________________

6.2 SQUARE ROOTS OF SUMS AND DIFFERENCES

In general there is no way to simplify sums and differences such as $\sqrt{x} + \sqrt{y}$. It is very tempting to separate the terms in a manner analogous to the way factors separate for multiplication, but such a separation is not correct. Specifically, $\sqrt{x} + \sqrt{y}$ is not equal to $\sqrt{x + y}$. To see this, consider the sum of square roots of two perfect squares.

$$\sqrt{4} + \sqrt{9} = 2 + 3 = 5,$$

whereas

$$\sqrt{4 + 9} = \sqrt{13} = 3.0555....$$

When dealing with sums and radicals, most often the only "simplification" that can be done is to factor common quantities. The expression

$$\sqrt{x^2y} + \sqrt{3y}$$

may be written

$$x\sqrt{y} + \sqrt{3y} + \sqrt{y}(x + \sqrt{3}).$$

This is as far as the expression may be changed. Whether or not it is simpler than the original is a matter of opinion.

Practice Problems

Simplify the following, if possible.

6. $\sqrt{2} + \sqrt{3}$ = ______________________

7. $\sqrt{7 + 5}$ = ______________________

8. $\sqrt{32} + \sqrt{2} =$ ______________________

9. $\sqrt{a^2b} + \sqrt{b^2a} =$ ______________________

10. $\sqrt{x^2y + y^2x} =$ ______________________

6.3 DIVISION INVOLVING RADICALS

Division by radicals is performed in the same manner as division by any other quantity. It is perfectly acceptable insofar as the mathematical meaning of an expression is concerned to leave a radical in any part of an expression. In fact, in some equations the denominator of a fraction may deliberately be left as a radical. However, for many applications you will want to rationalize the denominator. This means to convert a fraction having a radical in the denominator by multiplying numerator and denominator by the appropriate multiplier.

The fraction

$$\frac{1}{\sqrt{2}}$$

is not changed if top and bottom are multiplied by $\sqrt{2}$.

$$\frac{1}{\sqrt{2}} = \frac{1\ (\sqrt{2})}{\sqrt{2}\ (\sqrt{2})} .$$

The numerator of the product fraction is $\sqrt{2}$, and the denominator is the rational number 2. The new fraction is

$$\frac{\sqrt{2}}{2} .$$

More complex fractions are simplified in a similar manner, although the multiplier that rationalizes the denominator may not be as obvious as with the simple expressions. For example, the fraction

$$\frac{1}{1 - \sqrt{2}}$$

will have a rationalized denominator if multiplied top and bottom by

$1 + \sqrt{2}$.

$$\frac{1}{1 - \sqrt{2}} \cdot \frac{1 + \sqrt{2}}{1 + \sqrt{2}} = \frac{1 + \sqrt{2}}{(1 - \sqrt{2})(1 + \sqrt{2})} .$$

Since

$$(a + b)(a - b) = a^2 - b^2,$$

$$(1 - \sqrt{2})(1 + \sqrt{2}) = (1 - 2),$$

and the fraction becomes

$$\frac{1 + \sqrt{2}}{1 - 2} = -1 - \sqrt{2}.$$

Practice Problems

Simplify the following expressions.

11. $\dfrac{\sqrt{a^2b}}{\sqrt{b}}$ = ______________________

12. $\dfrac{\sqrt{2a^2b}}{\sqrt{ab}}$ = ______________________

13. $\dfrac{y\sqrt{x^2}}{\sqrt{y}}$ = ______________________

14. $\dfrac{A\sqrt{AB^2}}{\sqrt{A^2B}}$ = ______________________

15. $\dfrac{3pq}{\sqrt{2q}}$ = ______________________

Simplify the following by rationalizing the denominators.

16. $\dfrac{\sqrt{2 + 3}}{\sqrt{2} + \sqrt{3}}$ = ______________________

17. $\dfrac{\sqrt{a + b}}{\sqrt{a} + \sqrt{b}}$ = ______________________

18. $\dfrac{\sqrt{a^2 + b} + \sqrt{b}}{\sqrt{b^2 a} + \sqrt{a}}$ = ______________

19. $\dfrac{\sqrt{a^2 b} + a\sqrt{b}}{\sqrt{a} + \sqrt{b}}$ = ______________

20. $(\sqrt{x}\ \sqrt{x + y})/\sqrt{x^2 y}$ = _______

Answers to Practice Problems

1. $b\sqrt{6ab}$
2. $4y\sqrt{2x}$
3. $xy\sqrt{2y}$
4. $6zq^2$
5. $6A\sqrt{2Bxy}$
6. $\sqrt{2} + \sqrt{3}$
7. $2\sqrt{3}$
8. $5\sqrt{2}$
9. $a\sqrt{b} + b\sqrt{a}$
10. $\sqrt{x^2 y + y^2 x}$ or $\sqrt{xy}\ \sqrt{x + y}$
11. a
12. $\sqrt{2a}$
13. $x/\sqrt{y}$
14. $\sqrt{AB}$
15. $3p\sqrt{q}/\sqrt{2}$ or $3p\sqrt{2q}/2$
16. $-\sqrt{10} + \sqrt{15}$

17. $\dfrac{(\sqrt{a} - \sqrt{b})\sqrt{a + b}}{a - b}$

18. $\dfrac{b\sqrt{a^3 + ab} + b\sqrt{ab} - \sqrt{a^3 + ab} - \sqrt{ab}}{a(b^2 - 1)}$

19. $\dfrac{2a(\sqrt{ab} - b)}{a - b}$

20. $\dfrac{\sqrt{xy}\ \sqrt{x + y}}{xy}$

Exercises

1. $\sqrt{4}\,\sqrt{8}$ = ______________________

2. $\sqrt{27}$ = ______________________

3. $\sqrt{18}$ = ______________________

4. $\sqrt{50}$ = ______________________

5. $\sqrt{50}\,\sqrt{5}$ = ______________________

6. $\sqrt{a}\,\sqrt{b^2}$ = ______________________

7. $\sqrt{x^2}\,\sqrt{xy}$ = ______________________

8. $\sqrt{A}\,\sqrt{xA^3}$ = ______________________

9. $\sqrt{4\omega\theta}\,\sqrt{\omega^3}\,\sqrt{\alpha\theta\omega}$ = ______________________

10. $\sqrt{pg}\,\sqrt{2q^3}\,\sqrt{p^2q}$ = ______________________

11. $\sqrt{4} + \sqrt{9}$ = ______________________

12. $\sqrt{4 + 9}$ = ______________________

13. $\sqrt{3} + \sqrt{12}$ = ______________________

14. $\sqrt{a}\,\sqrt{a^3b}$ = ______________________

15. $\sqrt{x^2}\,\sqrt{xy^2}$ = ______________________

16. $\sqrt{p}\,(\sqrt{p^3} + \sqrt{p})$ = ______________________

17. $(\sqrt{A} + \sqrt{B})\,\sqrt{A^3B}$ = ______________________

18. $\sqrt{8a^3x}\,\sqrt{9ax^2}$ = ______________________

19. $\sqrt{4x^3y}/\sqrt{x^2y}$ = ______________________

20. $\dfrac{a\sqrt{b}}{3\,(\sqrt{b^3}-\sqrt{b})}$ = ______________________

21. $\dfrac{\sqrt{3}}{2 + \sqrt{3}}$ = ______________________

22. $\dfrac{\sqrt{2} + 2\sqrt{3}}{\sqrt{3} - \sqrt{2}}$ = ______________________

23. $\dfrac{\sqrt{x/y} - \sqrt{2xy}}{\sqrt{xy^3}}$ = ________________

24. $\dfrac{\sqrt{2p/3q} + \sqrt{q^3/p}}{\sqrt{6p^3}}$ = ________________

25. $\sqrt{8 + 3}\ \sqrt{8 - 3}$ = ________________

26. $\sqrt{a + b}\ \ \sqrt{a - b}$ = ________________

27. $\dfrac{\sqrt{2} + \sqrt{6}}{\sqrt{3} - \sqrt{2}}$ = ________________

7

Quadratic Equations

Quadratic equations were discussed in Chapter 5, although the equations of that chapter were not solved for any variables other than linear. As a brief review, any equation of the form

$$ax^2 + bx + c = 0,$$

where a, b, and c are constants and a is not equal to zero, is said to be a quadratic equation in the variable x. (In fact, this particular equation is often called the quadratic equation.) A quadratic equation is one where a variable appears to second order (that is, squared).

When the variable to be found in any equation is to first power (linear), the solution is straightforward and involves no special formula or rule other than maintaining the equality by always doing the same thing to both sides of the equation. When the variable being sought is the quadratic variable, it is no longer possible to isolate that variable by the simple operations of addition, subtraction, multiplication, and division. This may be seen from the defining equation

$$ax^2 + bx + c = 0.$$

The variable x may be isolated by subtracting c from each side,

$$ax^2 + bx = -c,$$

but it is impossible to go any further toward isolating x.

Instead of trying to isolate the quadratic variable by the basic mathematical operations, quadratic equations are solved by application of what is called the quadratic formula.

7.1 THE QUADRATIC FORMULA

Rule: Solutions to any quadratic equation of the form

$$ax^2 + bx + c,$$

where a, b, and c are constants, are

$$x = \frac{-b \pm \sqrt{b^2 - 4ac}}{2a}$$

The ± gives two solutions to the equation. These solutions are called roots. Any quadratic equation will always have two roots.

The radical in the quadratic formula may be positive, zero, or negative. When negative, the square root is said to be imaginary. Imaginary numbers are important to much of physics but are very seldom used in the introductory course. This workbook does not consider imaginary numbers, so all of the equations in this chapter will generate radicals that are zero or positive. Such roots are said to be real. They may very well, however, be irrational.

The quadratic formula will work for any equation of the correct form for any value of b or c (positive, negative, or zero), and for any value for a other than zero. When $a = 0$, the formula requires division by zero, which is not possible.

All equations do not have x as the variable and coefficients a, b, and c. These symbols stand for the multiplier of the squared and linear terms and the constant. Any quadratic equation must be put in the correct form before the formula may be applied. For example, consider the following equation.

$$2x(-2 + 4x) = 3x^2 + 4.$$

The first task is to clear the parentheses:

$$-4x + 8x^2 = 3x^2 + 4.$$

Next, put all terms on the same side of the equation:

$$-4x + 8x^2 - 3x^2 - 4 = 0.$$

Finally, arrange in descending powers of x:

$$5x^2 - 4x - 4 = 0.$$

Now the quadratic formula can be applied. In the quadratic formula, $a = 5$, $b = -4$, and $c = -4$:

$$x = \frac{-(-4) \pm \sqrt{(-4)^2 - 4(5)(-4)}}{2(5)} = \frac{4 \pm \sqrt{16 + 80}}{10},$$

$$= \frac{4 \pm \sqrt{96}}{10} = \frac{4 \pm 4\sqrt{6}}{10} = \frac{2 \pm 2\sqrt{6}}{5} \quad .$$

The individual roots to the equation are

$$x = \frac{2 + 2\sqrt{6}}{5} \text{ and } x = \frac{2 - 2\sqrt{6}}{5}.$$

Practice Problems

Solve for the variable in the following equations.

1. $-2 + 4x^2 - 8x = 0$

2. $3y + 2 = 6y(3 - y) + 5$

3. $-4(2a + 3) + 2a^2 = a(5 - a)$

Solve for x in the following equations.

4. $cx + bx^2 + a = 0$

5. $A(4 - x) + B = cx^2$

6. $(\Omega x - 4)(2x) = 3(x^2 + 2)$

7.2 FACTORING

When the quadratic equation has whole number roots, the radical $\sqrt{b^2 - 4ac}$ will be a perfect square. In these instances, it is possible to find the roots by a process called factoring. When the equation is factored, two terms are found which when multiplied will give the original equation. Since the product of the two terms is zero, if either one is zero the equality is satisfied. Each factor is set equal to zero and the solutions to the resulting linear equations give the roots to the equation. To see how this works, consider the equation

$$x^2 - 6x - 16 = 0.$$

The equation may be factored as

$$(x - 8)(x + 2) = 0.$$

The roots are given by the two conditions,

$$x - 8 = 0;\ x = 8$$

or

$$x + 2 = 0;\ x = -2.$$

Of course, the quadratic formula also gives the roots.

$$x = \frac{-(-6) \pm \sqrt{36 - (4)(-16)}}{2} = \frac{6 \pm \sqrt{100}}{2} = \frac{6 \pm 10}{2} = 3 \pm 5,$$

$$x = 8 \text{ or } x = -2.$$

Factoring is a skill that comes with practice. There is no quick and easy way to tell whether or not an equation can be factored. If the constants are large, the factoring may be complex even if the roots are rational, and it may be easier to apply the quadratic formula. However, for equations with small numbers for constants, it is usually possible to perform a quick test for factorability. As an example, consider the equation

$$3x^2 + 2x - 8 = 0.$$

If the equation is factorable, the factors will have leading terms that multiply to $3x^2$, so must be of the form

$$(3x\_\_\_\_\_\_\_)(x\_\_\_\_\_\_\_\_) = 0.$$

where the blanks represent terms to be determined.

The second terms in the factors multiply to give 8, and the only possible pairs are (8,1) and (2,4). There are only four possible pairings:

$$(3x \underline{?}\ 8)(x \underline{?}\ 1)$$

$$(3x \underline{?}\ 1)(x \underline{?}\ 8)$$

$$(3x \underline{?}\ 2)(x \underline{?}\ 4)$$

$$(3x \underline{?}\ 4)(x \underline{?}\ 2)$$

The ? represents the sign to be determined.

Since the constant in the quadratic equation is negative, the signs in the factors must be opposite. Taking the first set of pairs, two possibilities are to be tried,

$$(3x - 8)(x + 1) = 3x^2 - 5x - 8$$

and

$$(3x + 8)(x - 1) = 3x^2 + 5x - 8.$$

Neither of these yields the original equation. Moving on to the second pair,

$$(3x + 1)(x - 8) = 3x^2 - 23x - 8$$

and

$$(3x - 1)(x + 8) = 3x^2 + 23x - 8.$$

Again, neither is the correct equation. For the third pair we have

$$(3x + 2)(x - 4) = 3x^2 - 10x - 8$$

and

$$(3x - 2)(x + 4) = 3x^2 + 10x - 8.$$

Neither of these is correct. The final pair is

$$(3x - 4)(x + 2) = 3x^2 - 2x - 8$$

and

$$(3x + 4)(x - 2) = 3x^2 + 2x - 8.$$

The final pairing gives the correct equation, so the equation is written in factored form as

$$(3x + 4)(x - 2) = 0.$$

The roots are $x = 2$ and $x = -4/3$.

This step-by-step procedure may seem long when written out in all its detail, but in reality it is not as long as it seems. The possible roots 8 and 1 can be eliminated rather quickly, leaving only the 2,4 pair. With just a little practice, it becomes easy to recognize that the arrangement that multiplies 3x times 4 cannot possibly result in a coefficient of 2 for the middle term, so the final arrangement is reached after only at most a couple of tries.

Practice Problems

Solve for the indicated variable:

7. $8x^2 + 6x - 2 = 0$

8. $6z^2 + 7z - 3 = 0$

9. $a^2 - 4 = 0$

10. $2z^2 + 3z + 1 = 0$

Solve for x:

11. $a^2x^2 + 2abx + b^2 = 0$

Answers to Practice Problems

1. $\frac{2 + \sqrt{6}}{2}$ and $\frac{2 - \sqrt{6}}{2}$

2. $\frac{5 + \sqrt{33}}{4}$ and $\frac{5 - \sqrt{33}}{4}$

3. $\frac{13 \pm \sqrt{313}}{6}$

4. $\frac{-c \pm \sqrt{c^2 - 4ab}}{2b}$

5. $\frac{-A \pm \sqrt{A^2 + 4C\ (4A + B)}}{2C}$

6. $\frac{4 \pm \sqrt{12\Omega - 2}}{(2\Omega - 3)}$

7. $x = 1/4$ and $x = -1$

8. $z = -3/2$ and $z = 1/3$

9. $a = \pm 2$

10. $z = -1$ and $z = -1/2$

11. $x = -b/a$

Exercises

Solve for the indicated variable.

1. $2x^2 - 2x - 3 = 0$
2. $x^2 - 7x + 2 = 0$
3. $z^2 - 4z + 4 = 0$
4. $4y - 2 + y^2 = 0$
5. $3p + 3p^2 - 1 = 0$
6. $B - B^2 + 6 = 0$
7. $x^2 + 2 + 2x(x - 4) = 0$

8. $2(x^2 + 3) + 3(x - 4) = (x - 3)(x + 1)$

9. $(3q + 2)(q - 1) = q^2 - 4q$

10. $8s(s - 2) + 5(1 - s^2) = 5s(s + 1)$

11. $x^2 - x - 2 = 0$

12. $y^2 - y - 20 = y - 5$

13. $z^2 + 6 = 5z$

14. $x^2 + 6x + 8 = 2x + 5$

15. $x^2 + 6 = -(2 + 8x)$

16. $y^2 + 6y + 8 = 2x + 5$

17. $2A^2 + A - 1 = 0$

18. $3x^2 + 4x + 1 = 0$

19. $2x(x - 2) = 2(x + 4)$

20. $6a^2 + 7a - 3 = 0$

21. $8x^2 + 4x - 4 = 4x - 2$

22. $8x^2 + 4x = 2(1 - x)$

Solve for x in the equations below.

23. $Ax^2 + Bx + D = C$

24. $p(x^2 - 1) + qx = (p - q)x^2$

25. $(ax + b)(bx - a) = a - b$

26. $Dx^2 - w = \alpha x(D + 2)$

27. $a(x^2 - 1) - b(x + a) = (x - a)(x + a)$

28. $ax(cx + b) = -a(bx + c) + bx$

29. $p = p_0 + v_0 x + (1/2)ax^2$

30. $2x(x + y) + x(x - y) = y$

Solve for t in the equations below.

31. $x = x_0 - v_0 t + (1/2)at^2$

32. $Dt^2 - Ct = D(C - D)$

33. $t^2/3 - 12t/7 = 8$

34. $2t^2 - 4t - 2 = 0$

35. $2t^2 + 3t - 2 = 0$

36. $2t^2 - 2t - 2 = 0$

37. $2t^2 - 4t + 2 = 0$

38. $At^2 - 2At + A = 0$

39. $xyt^2 + 2xyt + xy = 0$

40. $A^2t^2 + ABt = CB + ACt$

8

Equations in Two Variables

Although the equations in the preceding chapters often contained several symbols, only one was treated as "the variable." The intent was always to find the value of this single variable or, in the instance where other symbols representing constants were present, an expression for this variable in terms of the other symbols. In many instances, a solution to a physics problem will require introducing more than one equation with more than one variable. This may seem strange, but often it is extremely difficult, if not impossible, to measure directly the desired quantity. However, if the desired quantity can be calculated from the values of other quantities which are directly measurable, then the desired quantity can be determined. Such a circumstance is quite common in physics problems and laboratories.

There is one limitation on situations with multiple variables. For a given set of unknowns, the number of independent equations containing those unknowns must be exactly the same as the number of unknowns.

> Rule: When a variety of conditions are imposed in a situation from which equations are written, there must be exactly the same number of independent equations as unknowns in order to have a unique solution for the unknowns.

An equation is independent if it comes from a set of conditions different from the conditions which gave any of the other equations. An independent equation cannot be obtained by manipulation of existing equations.

In principle there is no limit to the number of unknowns and equations that might be involved in a given problem. However, most beginning physics courses will not require you to work with more than two equations and two unknowns. In this chapter only two equations with two unknowns will be considered.

8.1 LINEAR EQUATIONS IN TWO UNKNOWNS

Linear equations were introduced earlier as equations containing variables to first power only. If two equations and two unknowns are given and both are linear, then both contain the two variables to only first power.

> Rule: Given two linear equations in two unknowns, eliminate one of the unknowns by manipulation of the equations. Solve for the remaining unknown from the resulting equation.

To see how the rule is applied, consider the pair of equations below.

$$x + y = 4,$$

$$x - y = 6.$$

If these equations are added, the result is an equation in x only.

$$\begin{array}{rl} & x + y = 4 \\ + & \underline{x - y = 6} \\ & 2x \quad = 10 \\ & \quad x = 5. \end{array}$$

The value of y is found by substituting $x = 5$ into either of the original equations, with the result $y = -1$.

Directly adding or subtracting the equations will not always eliminate one of the unknowns. Consider the equations

$$3p + 2q = 4,$$

$$-2p + q = 1.$$

Adding the equations as they are given gives a third equation containing both p and q. However, if the second equation is multiplied by 2 and then subtracted from the first, the resulting equation will contain only p:

$$\begin{array}{ll} 3p + 2q = 4 & 3p + 2q = 4 \\ (2)(-2p + q = 1) & -4p + 2q = 2 \end{array}$$

Subtracting the second from the first gives

$$7p = 2,$$

$$p = 2/7.$$

Thus, $p = 2/7$ satisfies both of the original equations. q is found by substituting this value for p into either of the original equations, giving $q = 11/7$.

The values of p and q could have been found by an alternative approach. If one variable is found in terms of the other from one equation, and this substituted into the remaining equation, the resulting equation will contain only one unknown. This unknown is then found by the methods of Chapter 5. To see how this works, consider the following equations.

$$3x - 2y = 4,$$

$$2x - 4y = 3.$$

Solve for x in the first equation, $3x - 2y = 4$,

$$3x = 4 + 2y,$$

$$x = (4 + 2y)/3.$$

Substitute this value for x into the second equation, $2x - 4y = 3$:

$$2(4 + 2y)/3 - 4y = 3.$$

Multiply by 3:

$$2(4 + 2y) - 12y = 9,$$

$$8 + 4y - 12y = 9,$$

$$8 - 8y = 9,$$

$$-8y = 1,$$

$$y = -1/8.$$

Substituting this value for y into the second equation gives:

$$2x - (4)(-1/8) = 3$$

$$2x + 1/2 = 3$$

$$2x = 5/2$$

$$x = 5/4.$$

The solution is $x = 5/4$, $y = -1/8$.

We have used two methods to solve two equations in two unknowns. The method you choose to use is a matter of personal choice. The solution is unique.

Practice Problems

Solve for the variables:

1. $2x - y = -1$
   $3x = y$

2. $3p - 4q = 2$
   $p + 2q = -3$

3. $-4A + B = -2$
   $3A + 2B = 1$

4. $-t + 3r = 8$
   $4t - r = 2$

8.2 WORD PROBLEMS

Many students find word problems difficult. Word problems require several steps, any one of which is necessary and none of which is sufficient. A breakdown at any of the steps means the problem will not be solved. Earlier in this workbook word problems were discussed, but until this point the problems have been relatively simple because only one unknown was involved. The logic required to write out an equation was correspondingly simple. With two variables, the logic as well as the solution becomes more difficult.

The first step in solving any problem is to be sure that you understand the problem. With a word problem, you must read the material for content, translate what you have read into an equation or set of equations, and finally apply the rules of algebra to find the desired solution. This reading step is extremely important, because often extraneous material is included that really has nothing to do with the problem. Unless you are clear about what material is important and how it should be used, there can be no hope for a solution.

As an example of how a word problem is approached, consider the following problem.

> The sum of two numbers is 33 and their difference is 9. What are the two numbers?

Two numbers are to be found, so let us begin by calling the numbers x and y. We are told that the sum of the numbers is 33, and the sum of the numbers is written symbolically as $(x + y)$, so we may write the equation

$$x + y = 33.$$

The other information is that the difference between the numbers is 9. The difference is $(x - y)$, so the second equation is

$$x - y = 9.$$

The two equations to solve are

$$x + y = 33,$$

$$x - y = 9.$$

Adding the equations yields

$$2x = 42,$$

$$x = 21.$$

Substituting y into either equation, we find that $y = 12$. To check:

$21 + 12 = 33$ and $21 - 12 = 9$.

As a second example, consider the following:

> If three times the larger of two numbers is added to two times the smaller, the result is 46. If three times the smaller is subtracted from five times the larger, the result is 26. What are the numbers?

Again we are looking for two numbers, and again we will let the numbers be x and y, where x is the larger and y the smaller. The statement "...three times the larger...is added to two times the smaller..." becomes $3x + 2y$, and this is given as equal to 46.

$$3x + 2y = 46.$$

The second statement, "...three times the smaller is subtracted from five times the larger..." is written $5x - 3y$, and this is equal to 26.

$$5x - 3y = 26.$$

The task now is to solve the equations

$$3x + 2y = 46,$$

$$5x - 3y = 26.$$

One method of solution is to multiply the first equation by 3 and the second equation by 2 and add.

$$\begin{array}{rl} 9x + 6y &= 138 \\ + \quad 10x - 6y &= 52 \\ \hline 19x &= 190, \end{array}$$

from which $x = 10$. Putting this value of x into either of the original equations gives $y = 8$. To check: $3(10) + 2(8) = 46$ and $5(10) - 3(8) = 26$.

Practice Problems

5. The sum of two numbers is 13 and their difference is 1. What is the number?

6. Bill is two years older than twice as old as his daughter. Together their ages add to 62. How old are Bill and his daughter?

7. One day Betty counted her money. She had $15.00 in bills and $1.55 in change. Her bills were a mixture of one dollar bills and five dollar bills, and her change was all in nickels and dimes. She had seven more nickels than dimes. How many nickels and dimes did she have?

8. The perimeter of a rectangular plot is 150 feet. The length of the plot is twice the width. What are the dimensions of the plot?

9. Four bars of candy A and 2 bars of candy B cost $1.70. Two bars of candy A and four bars of candy B cost $1.90. What is the cost of each type of candy?

8.3 EQUATIONS IN TWO VARIABLES, ONE A QUADRATIC

Aside from trigonometric functions (discussed in Chapter 10), the most complex function you are likely to encounter in an introductory physics course is a quadratic (i.e., the highest power of a variable is the square). When faced with a set of parametric equations where quadratics are involved, the procedure is parallel to that when the equations were linear. By some process, eleminate one unknown so that you are left with one equation in one unknown and then solve for that unknown. Substitute back into either original equation and find the second unknown. Most of the problems involving a quadratic that you will have to solve will have one linear equation and one quadratic. For these, it is usually simpler to solve the linear for one unknown and substitute that into the quadratic.

A quadratic equation always has two roots. These are a result of the algebraic solution, not necessarily the physics. In other words, your algebra may generate two possible solutions, only one of which actually satisfies the physics, and so you will have to interpret the answers to determine which makes sense physically. This may seem strange, but the equations of physics are written for a range of values for variables. In that range certain conditions apply. If the conditions change, the equations that describe a phenomenon also change, and the original set of equations is not valid. Even so, the original equations still have solutions, although these solutions do not represent physical reality. You will encounter one example of this in the study of motion. Imagine you are stopped at a stop sign and at the moment you move away from the sign someone starts a stopwatch. Your position in time may readily be expressed as an equation in variables representing position and time. If you substitute negative values for time into this equation, the equation will describe that you were at some negative value of position while in fact you were parked at the sign. For the negative values of time, you were not moving and the equations used for positive time simply are not valid. The equation may be solved, but it doesn't mean anything. You must always examine your solutions to check that they in fact answer the question you set out to answer.

As an example of how to solve two equations in two unknowns, one a quadratic, consider the pair below.

$$2x - y = 2,$$

$$x^2 - 2y^2 = -12.$$

From the first,

$$y = 2x - 2.$$

Substituting this into the second gives

$$x^2 - 2(2x - 2)^2 = -12,$$

$$x^2 - 2(4x^2 - 4x + 4) = -12$$

$$x^2 - 8x^2 + 8x - 8 = -12$$

$$-7x^2 + 8x + 4 = 0$$

Application of the quadratic formula gives

$$x = \frac{-16 \pm \sqrt{(16)^2 - (4)(-7)(4)}}{2(-7)} = \frac{-16 \pm \sqrt{368}}{-14} = \frac{8 \pm 2\sqrt{23}}{7}.$$

This is as far as the solution for x can be reduced. If appropriate, the value for $\sqrt{23}$ could be substituted and numerical values determined for x, rounded off appropriately. y is found by substituting the values for x into either equation, although the linear is by far the easier to work with. The result is

$$y = \frac{2 \pm 4\sqrt{23}}{7}$$

As a second example, consider the equations below.

$$-x + 2y = 4$$

$$2x^2 + 4y^2 = 12.$$

Solving for x in the first equation gives

$$x = 2y - 4.$$

Substituting this into the second equation yields

$$2(2y - 4)^2 + 4y^2 = 12,$$

$$2(4y^2 - 16y + 16) + 4y^2 = 12,$$

$$8y^2 - 32y + 32 + 4y^2 = 12,$$

$$12y^2 - 32y + 20 = 0,$$

$$3y^2 - 8y + 5 = 0,$$

$$(3y - 5)(y - 1) = 0,$$

$$y = 5/3 \text{ and } y = 1.$$

The pair of solutions is obtained by substituting each value for y into one of the original equations, with the resulting pair of solutions

$$(y = 5/3,\ x = -2/3),$$

$$(y = 1,\ x = -2).$$

Practice Problems

10. $x = 2y - 3$

    $x^2 + 2y^2 = 9$

11. $2x + y = 8$

    $2x^2 + y^2 = 54$

12. $-3x + 6y = 9$

    $x^2 + y^2 = 2$

13. $x + y = 3$

    $2x^2 - y^2 = 4$

14. $-2x + 5y = -2$

    $2x^2 + 5y^2 = 2$

Answers to Practice Problems

1. $x = 1, y = 3$
2. $p = -4/5, q = -11/10$
3. $A = 5/11, B = -2/11$
4. $r = 34/11, t = 14/11$
5. 6 and 7
6. 42 and 20
7. 8 dimes and 15 nickels
8. 25 by 50 feet
9. Bar A cost 25 cents and bar B cost 35 cents.
10. $(y = 0, x = -3)$ and $(y = 2, x = 1)$
11. $(x = 1/3, y = 23/3)$ and $(x = 5, y = -2)$
12. $(y = 7/5, x = -1/5)$ and $(y = 1, x = 1)$
13. $(y = 6 + \sqrt{22}, x = -3 - \sqrt{22})$ and $y = 6 - \sqrt{22}, x = -3 + \sqrt{22})$
14. $(x = -3/7, y = -4/7)$ and $(x = 1, y = 0)$

Exercises

Solve for the variables.

1. $$\begin{aligned} 2x - y &= 4 \\ x &= 2y \end{aligned}$$

2. $$\begin{aligned} x - 2y &= 4 \\ x + 2y &= 1 \end{aligned}$$

3. $$\begin{aligned} 2x + 2y &= 5 \\ -x + 3y &= 2 \end{aligned}$$

4. $$\begin{aligned} -3p + 2q &= 4 \\ 5p - q &= 1 \end{aligned}$$

5. $$\begin{aligned} 2a - 4b &= 5 \\ 3a + 4b &= 2 \end{aligned}$$

6. $$\begin{aligned} 4A - C &= 8 \\ 3A &= 2C - 2 \end{aligned}$$

7. $$\begin{aligned} 2x - 8 &= 3y \\ 8 + 2y &= 5x \end{aligned}$$

8. $3x + 2y = 10$
$2x = y + 8$

9. $-7p + 13 = 2b$
$4b + 2p = 6$

10. $3x - 8y = 0$
$4x - y = 0$

11. $x - 2y = 4$
$2y = -x - 3$

12. $2y = 4 + p$
$8 - p = -y$

13. $3z - 8 = 4A$
$A + 6z = -2$

14. $-5x - 7y = 9$
$3x - 2y = 0$

15. $3x - 6y = 12$
$4x = 24 - 8y$

16. $2x + y = -7$

$4x^2 + 2y^2 = 66$

17. $-6x + 8y = 2$

$3x^2 + 4y^2 = 3$

18. $3v_1 + 2v_2 = -4$

$3v_1^2 + 2v_2^2 = 14$

19. $3p + 8q = 14$

$3p^2 + 8q^2 = 20$

20. $p + 4q = -1$

$p^2 + 4q^2 = 29$

21. $x^2 = 2t^2$

$2t = x - 4$

22. $6a - 2 = b$

$3a^2 = b^2 - 1$

23. $6A - B = 4$

$A^2 + 4 = B^2$

24. $3p - 6 = 9q$

$p^2 + 3q^2 = 14$

25. A class of 30 students has twice as many girls as boys. How many boys and girls are in the class?

26. The sum of two numbers is 15. The larger number is four times the smaller. What are the two numbers?

27. Two cars start out from the same point and drive along the same highway. Car A travels at a steady 55 miles per hour and car B travels at a steady speed of 60 miles per hour. How far apart are the cars after driving four hours?

28. Lucy bought five cans of tomato soup and three cans of bean soup. When her roommate asked her what each cost, she had forgotten. However, she did remember that she spent a total of $1.85 and that the tomato soup cost 5 cents a can more than the bean soup. How much did each type of soup cost?

29. Today Joe is twice as old as Maybelle and five years ago he was four years older than three times her age at that time. What are their ages today?

30. Frank's bank contains 27 coins, dimes and quarters only. The total amount of money in the bank is $4.50. How many dimes and how many quarters are in the bank?

31. The sum of two numbers is 21. The larger is one more than three times smaller. What are the two numbers?

32. A rectangle has a total perimeter of 10 inches. Twice the length is equal to three times the width. What are the dimensions?

33. Mr. Jones sells apples and oranges by the basket. Each basket sells for $6.00 and contains 18 pieces of fruit. Mr. Jones wants to get 30 cents for each apple and 35 cents for each orange. How many apples and oranges are in each basket?

34. The sum of two numbers is 30. The larger is 9 less than twice the smaller. What are the two numbers?

9 Parametric Equations

Very often the laws of physics are applied independently to parts of a situation as if the different parts are independent of each other. When this happens, there will always be common variables for each part. Each of these common variables is called a parameter.

For example, when a ballplayer throws a ball to home plate, the ball moves both vertically and horizontally. The laws of motion depend on three variables: acceleration, initial velocity, and time. Of these three variables, the vertical and horizontal components of acceleration and initial velocity are independent. In describing the motion of the ball, it is very convenient to write the equations for the vertical and horizontal motion independent of each other, each involving its own acceleration and initial velocity and each using a common parameter (or variable), time. This approach generates two equations for velocity and two equations for position, each containing the common variable time. The pair of equations for velocity and the pair of equations for position are each called parametric equations because each pair is coupled through the common parameter time.

The functional relation between the variables that were treated as independent of each other (but dependent on the common parameter) may be found by eliminating the parameter, leaving only the original variables that were treated originally as independent. In the example of the thrown ball, the ballplayer wants the ball to arrive at the catcher's mitt, imposing a very specific requirement on the vertical and horizontal positions at home plate. In the sense that he also wants the ball there before a runner, time is important. However, a wide range in time is acceptable (the ball can arrive one second or several seconds before the runner). The specific condition most important is that the ball and glove be at the correct vertical and correct horizontal position at exactly the same time, any time before the runner arrives. The answer to the question "Will the ball strike the catcher's mitt?" is contained in the equation between vertical and horizontal positions, time not appearing in the equation. The two equations for vertical and horizontal position may be reduced to one equation by eliminating time.

This process may be understood from a purely mathematical analysis. Using the proper relation between acceleration, initial velocity, and time, two equations in vertical position and horizontal position may be obtained, each containing time as a variable. This gives two equations in three unknowns. Since the number of independent variables does not match the number of equations, it is impossible to solve for unique values of any of the three. However, it is possible to reduce the number of unknowns to two by solving for the parametric variable in one and substituting into the other equation. This process leaves one equation in two unknowns. In the case of the thrown ball, the two remaining variables are vertical position and horizontal position. The equation is the equation for the path of the ball across the playing field. The path is a curve in space, independent of time.

In the beginning physics course, it is very unusual to find more than two parametric equations in a given situation. In this workbook we will not be concerned with more than two equations in two unknowns and a third common variable, which is the parameter.

9.1 LINEAR PARAMETRIC EQUATIONS

The procedure for eliminating a parameter between equations is very simple. Identify the parameter as the common variable in both equations, solve for this variable in one equation and substitute into the other. To see the procedure, consider the following equations.

$$3x + 2y = 4,$$

$$2x = 6 - z.$$

The common parameter is x (it appears in both equations), so the procedure is to eliminate x. From the second equation,

$$x = \frac{(6 - z)}{2}.$$

Substituting this into the first:

$$\frac{(3)(6 - z)}{2} + 2 = 4,$$

$$18 - 3z + 6y = 8$$

This equation gives the relation between z and y, independent of x. Of course, it should be simplified, and it may be that one variable is being sought in terms of another. Since we are only looking for the equation relating z and y, we will leave it as

$$6y + 10 = 3z.$$

From this equation, we could solve for y as a function of z or z as a function of y.

Practice Problems

Eliminate the parameter in the equations below and find an equation relating the remaining variables.

1. $x = 3t + 6$
   $y = 2t$

2. $2x = 3q - 2$
   $3p - 2 = x + 2$

3. $4 = 6x - p$
   $q - p = 4$

4. $7x - y = y + 2$
   $2a + 4 = y - 1$

In the equations below, uppercase (capital) letters represent constants and lowercase letters variables. The parameter is t. Eliminate the parameter between the equations and find an equation relating the remaining variables.

5. $Ax - B = At$
   $By - At = A - B$

6. $a = Dt + C$
   $b = Ct - D$

7. $Ap = At - C$
   $Bq = At + d$

8. $x(A - Bt) = Ct$
   $By - (A - t) = Ct$

9.2 PARAMETRIC EQUATIONS, ONE A QUADRATIC

When one of the parametric equations is a quadratic and the other linear, solution is exactly the same as when the equations are both linear. Solve for the parameter in one equation and substitute into the other. Although in general it is easier to solve for the parameter in the linear equation and substitute into the quadratic, the result is the same regardless of which equation is chosen to solve for the parameter for substitution into the other.

As an example, consider the following equations.

$$3x = t - 7,$$

$$y + t^2 = 2t + 1.$$

The parameter is t, and from the first equation,

$$t = 3x + 7.$$

Substituting this into the second equation gives

$$y + (3x + 7)^2 = 2(3x + 7) + 1,$$

$$y + 9x^2 + 42x + 49 = 6x + 14 + 1,$$

$$y + 9x^2 + 36x + 34 = 0.$$

This is the equation relating the variables x and y without parameter t. If the goal is to solve for y,

$$y = -9x^2 - 36x - 34.$$

If the goal is to solve for x, the solution is a little more complex. First arrange into the quadratic form:

$$9x^2 + 36x + (34 + y) = 0.$$

Note that y is treated as a constant because the quadratic equation can only solve for one variable at a time.

Apply the formula:

$$x = \frac{-36 \pm \sqrt{(36)^2 - (4)(19)(34 + y)}}{(2)(9)} = \frac{-36 \pm \sqrt{1296 - 1224 - 36y}}{18} =$$

$$x = \frac{-36 \pm \sqrt{72 - 36y}}{18} = \text{(eventually)} \ \frac{-6 \pm \sqrt{2 - y}}{3}$$

Practice Problems

Identify the parameter in the equations below and solve for the equation relating the other two variables.

9. $3p = t$

   $q^2 = p + 4$

10. $x = t^2$

   $y - t = 3$

11. $2a - 2 = b$

   $2(c + a) = 3c^2 - 1$

Solve for x as a function of v by eliminating the parameter t.

12. $v = 8t + 3$

   $x = -2 + 3t - 4t^2$

13. $3(t - 2) = 4v + 2t^2$

   $7(1 - t) - 6 = 2x$

In the problems below, consider capital letters as constants and lowercase letters as variables. Eliminate the parameter and find the equation relating the other variables.

14. $t = Ay^2$

   $a = By + A$

15. $c = Bx - D$

   $d - x = Dx\ (A - x)$

In the following equations, eliminate t as a parameter and solve for x as a function of y. Consider capital letters as constants.

16. $y = At^2$

   $x = Bt + A$

17. $y = Bt - D$

$x - t = Dt(A - t)$

18. $Ax = Bt + C$

$By = At^2 + B$

Answers to Practice Problems

1. $2x = 3y + 12$
2. $6p = 3q + 6$
3. $q = 6x$
4. $7x = 4a + 12$
5. $By - Ax = A - 2B$
6. $bD = aC - C^2 - D^2$
7. $Bq = Ap + C + D$
8. $ACx - Ax - B^2xy + ABx = CBy - CA$
9. $3q^2 = t + 12$
10. $x = (y - 3)^2$
11. $3c^2 - 2c = b + 3$
12. $16x = -v^2 + 12v - 59$
13. $8x^2 + 34x + 275 = -196v$
14. $B^2t = Aa^2 - 2aA^2 + A^3$
15. $B^2d - BD = ADBC - 2D^2c - Dc^2 + AD^2B - D^3 + CB$
16. $x = \dfrac{2A^2 \pm \sqrt{4A^4 - 4A(A^3 - B^2y)}}{2A}$
17. $x = \dfrac{-Dy^2 + y(ABD + B - 2D^2) + ABD^2 - D^3 + BD}{B^2}$
18. $x = \dfrac{2AC \pm \sqrt{4A^2C^2 - 4A(AC^2 + B^3 - B^3y)}}{2A^2}$

Exercises

Identify the parameter and eliminate it to obtain one equation relating the remaining variables. Treat uppercase (capital) letters as constants.

1. $3t - 4 = 5x$

$4y + 2 = t$

2. $4x - 3 = 6t$

$3y - x = 4$

3. $6(a - 2b) = 4 - a$

$6d - 3a = a + 1$

4. $2x - 3 = x\ (2 + p)$

$3y(2 + p) = py + 1$

5. $2Az - t = A$

$w = t + 2$

6. $Vt - 4 = x$

$y + 2 = Ut$

7. $R(x - 2) = 3y + B$

$Xz + y = BR$

8. $Ba(A - b) = CD$

$cb = D(Ab + cD)$

9. $2R(a - 4Bp) = 3a$

$Zb(p - A) = b + 1$

10. $BAx + 2 = 3t$

$B(y - 4) = Ayt$

11. $x = V_ot + (1/2)At^2$

$v = At + V_o$

12. $3a - b = b^2$

$a(c - 2) = 4c$

13. $v = aT$

$x = V_0T + (1/2)aT^2$

14. $2(p - q) = qp^2$

$3p - 4\Omega = \Omega$

15. $Ax - By = C$

$Bz - Cy^2 = A$

16. $Ax = Bt^2 + C$

$y = -Bt + 7$

17. $a = Bt + Ca$

$D(t^2 - b) = Cbt$

18. $Rr + S = St$

$Rs - St = T$

19. $Bb - Ct = t^2$

$Aa = (B - C)t$

20. $a = B + t/A$

$Bt = T^2b + B/A$

Solve for x as a function of y.

21. $x = 4t$

$y = 3t^2 - 4$

22. $2t + 3y = 4$

$2x^2 + 3t = 8x$

10

Graphs

Graphs are widely used in physics to present a visual representation of the interdependence between variables. Although it is possible to make perspective drawings representing three dimensions, the usual practice is to restrict graphs to two variables.

The variables plotted on a graph may represent any physical quantity. In all probability, the first graphs a student of physics encounters will be plots of some parameters of motion against time. We can't "see" time, but we can use an axis to represent changes in time. A graph plotting time versus some physical property is a very convenient mechanism for visualizing how the physical property changes in time. Such a plot can reveal at a glance such information as when the property has maximum or minimum values, when it is changing rapidly or slowly, and when the property is positive or negative. Of course any information is contained in equations that relate the physical property to time, but often the equations are not so easily interpreted. Even worse, it is not always possible to write the physical property in a nice, easily handled and understood equation. A graph may be drawn from measured data, whether or not the "equation" is known.

When graphs are studied in the context of a mathematics course, it is common practice to represent the dependent variable on the vertical axis and to represent the independent variable on the horizontal axis. In physics, the important thing is usually the interdependence of two variables, and the words "dependent" and "independent" may or may not be used.

The word function is often used, as in the statement "y is a function of time." In this context, the implication is that there is an equation relating y and time, and for the purposes under discussion it is of interest to express y as a function of time in the form y = whatever function. Just as often, it is of interest to know when something has a given value for y, and the relationship is then manipulated to solve for t, giving an equation t = whatever function. The equation has not changed; it has been rearranged. It may be said that y is the dependent variable in the first case and t the dependent variable in the second, although most physics teachers would probably not get too excited about the distinction.

There are some traditions in the graphs of physics. These are more for familiarity than deep meaning. For example, time is usually taken to be along the horizontal axis, positive to the right, and negative to the left. (Time in physics always means clock time, not absolute time, and as such may be positive or negative depending on whether before or after what is chosen to be called t = zero.) Spatial coordinates are traditionally plotted as the horizontal coordinate positive to the right along the horizontal axis and the vertical coordinate positive up along the vertical axis. In no case is the physics changed according to how a graph is plotted.

Graphs are not very accurate. When working with graphs to obtain numerical values, it is especially important to follow the rules of significant figures.

10.1 LINEAR GRAPHS

A linear graph is, as the name implies, a straight line. That is, if two variables have a linear relation, a plot of one against the other will be a straight line. Many textbooks use the equation

$$y = mx + b$$

as the "standard" equation for a linear relationship, where y becomes the variable plotted on the vertical axis and x the variable on the horizontal axis. The quantities m and b represent constants, respectively called the <u>slope</u> and the <u>intercept</u>. For purposes of explanation, we will examine the <u>graph</u> of the curve $p = 2q - 3$ in detail.

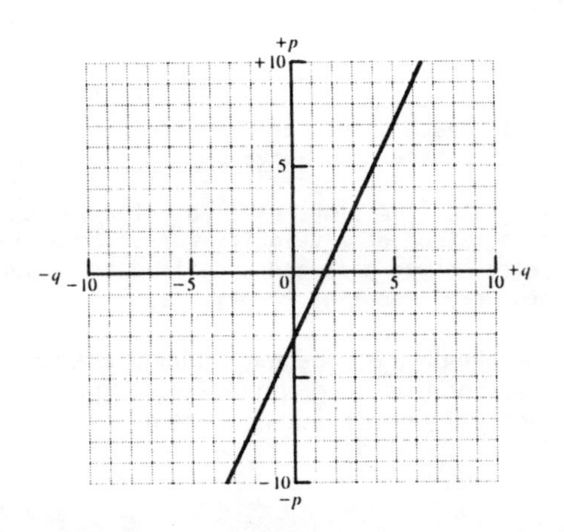

An equation represents the relationship between the variables of the equation for all possible values of the variables, ranging from + ∞ ("plus infinity") to - ∞ ("minus infinity"). On any finite page, only a portion of the graph may be represented and the graph plotted covers only a small range of the values of the variables. In our example graph, the region around the origin is displayed.

The meaning of the intercept (-3 in this example) may be seen from the graph. The <u>intercept</u> is the value of the variable on the vertical axis when the variable on the horizontal axis is zero. When $q = 0$, $p = -3$. You can see this from the graph or may determine it from the equation $p = 2q - 3$. The point where the line crosses the horizontal axis (at $q = 3/2$) does not have a special name.

The other parameter of the line is the multiplier of the horizontal variable (q in this example). This is called the <u>slope</u>, and is representative of the angle the line makes with the horizontal axis (the "steepness") of the line).

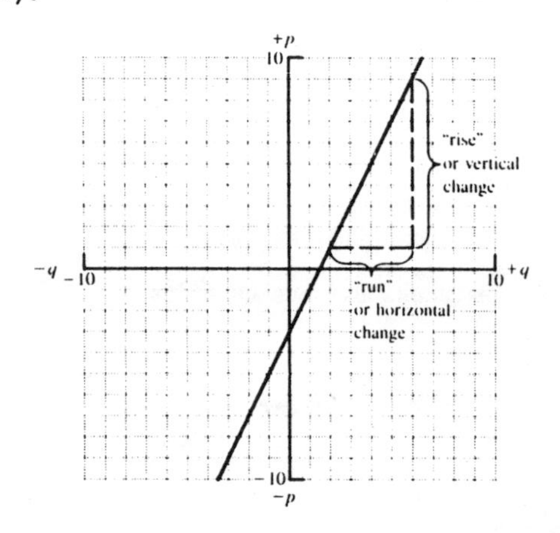

Sometimes the slope is defined as "rise over run." This phrase is a simplification of the general definition: The slope is the ratio of change in vertical dimension to change in horizontal dimension. In the example being considered, a slope of 2 means that the line increases by two units vertically for every increase of one unit horizontally. The "rise over run" phrase came into practice when surveyors were describing the slope of a hill, but the procedure is applicable to any linear graph, even when the variables are not spatial. The slope of a position-versus-time graph is called <u>velocity</u> and the slope of a velocity-versus-time graph is called <u>acceleration.</u> These and other parameters defined as a slope of the appropriate curve are very important to physics.

The slope of a straight line is constant. It makes absolutely no difference where the slope is determined. In the graph representing $p = 2q - 3$, both p and q are negative to the left of the p-axis (that is, in the region where q is less than 0). The slope is still 2.

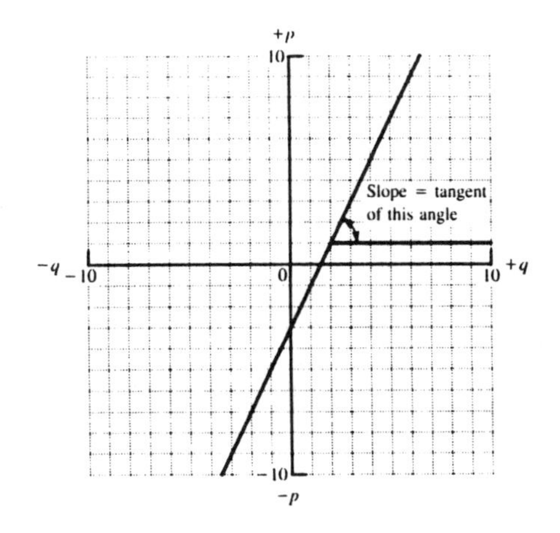

The slope may also be defined in terms of the angle the line makes with the horizontal axis (and of course with any line parallel to the horizontal axis, as in the graph above). If a right triangle is drawn using the line segment as hypotenuse and the vertical and horizontal changes as sides, one of the angles of the triangle will be the angle between the line and horizontal direction. From trigonometry, the ratio of the side opposite an angle in a right triangle (other than the right angle) to the side adjacent (not the hypotenuse) is called the tangent of the angle. In the little triangle which has the line as hypotenuse, the ratio of the vertical change to horizontal change gives the tangent of the angle between the line and a horizontal line. The slope of the line is the tangent of the angle made by the line with the horizontal direction. Since all right triangles drawn with line as hypotenuse and vertical and horizontal directions as sides will be similar to all triangles drawn using the same conditions, the ratios of the corresponding sides will be the same for every one of such triangles. The slope of the line is constant.

Slopes may be converted directly to angle between line and horizontal direction. A slope of one is the same as a slope of 45°, slope of 2 to 63.4°, slope zero to 0° or 180°, and so on. You will probably encounter slopes referred to interchangeably as degrees or as numbers.

A slope may also be negative, corresponding to a negative change in vertical variable for positive change in the horizontal variable.

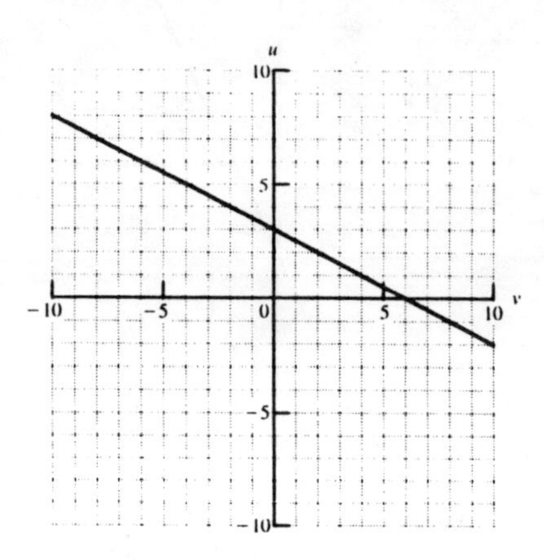

The negative also changes the angle when expressed in degrees, a negative angle being measured clockwise. Thus, -45° is an angle of 45° measured down from the horizontal.

To calculate the slope of the line in the graph above, pick two convenient points on the line and measure the corresponding changes in vertical and horizontal directions for those points. Your measurement will be more accurate the farther apart the points are located. Since inaccuracy in reading the graph is not so much percentage as a fixed amount (for example, one tenth of a division, or one fourth of a measurement), the smaller that fixed amount is compared to the total measurement, the smaller the percentage error. For the two points marked A and B, the vertical changes by -9 for a change of +18 in the horizontal, so the slope is -9/18 = -1/2 or -26.5°.

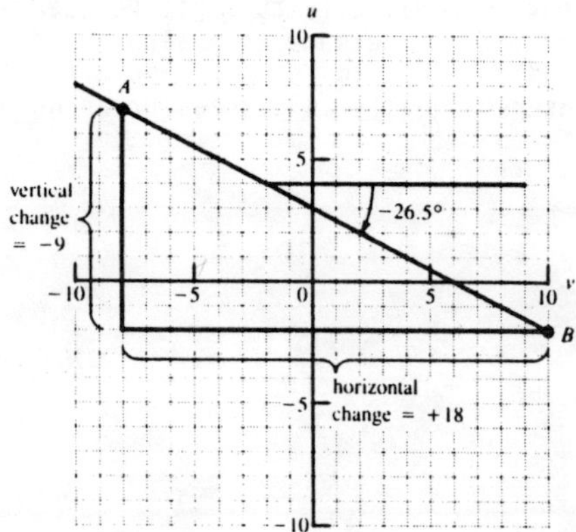

It is also possible to write the equation of this line. We only need put it in the form $y = mx + b$, where y is the variable on the vertical axis, x the variable on the horizontal axis, m the slope, and b the intercept. Reading from the graph, the intercept is +3. We have already determined the slope to be -1/2, so the equation of the line is

$$u = (-1/2)v + 3.$$

When the region displayed does not include that portion of the vertical axis where the line crosses the axis, the intercept cannot be read directly. The equation of the line can, however, be determined from any two points.

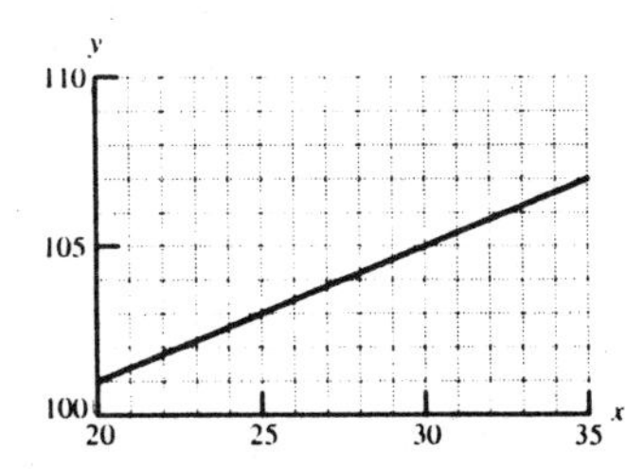

In the line illustrated, $y = 101$ when $x = 20$ and $y = 107$ when $x = 35$. Substituting into the equation

$$y = mx + b,$$

we have two equations in the two unknowns m and b:

$$101 = 20m + b,$$

$$107 = 35m + b.$$

Using techniques from preceding chapters, these give $m = 0.4$ and $b = 93$.

Practice Problems

Write the equation for each of the graphs below.

1.

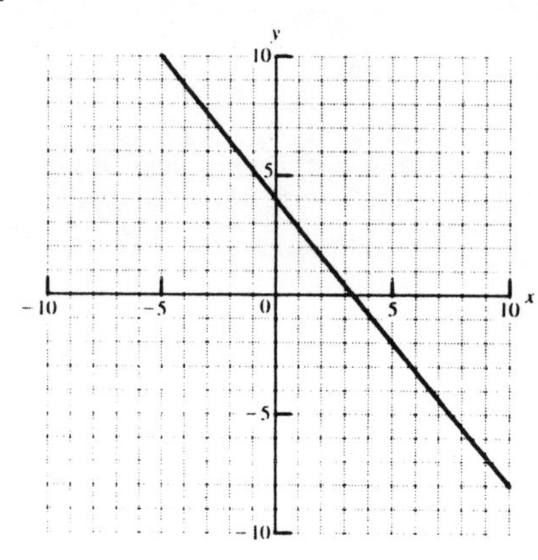

2.

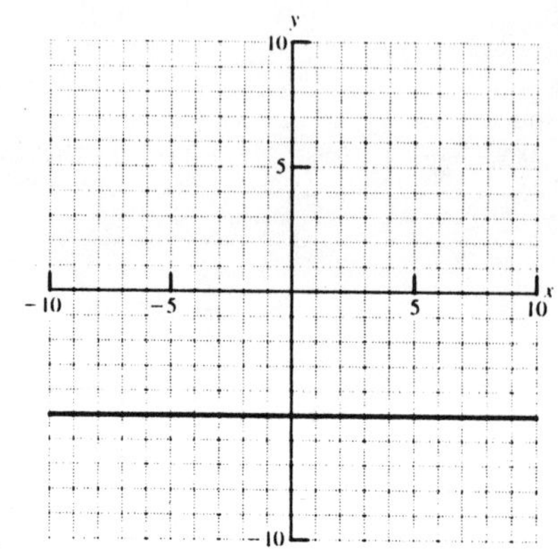

3.

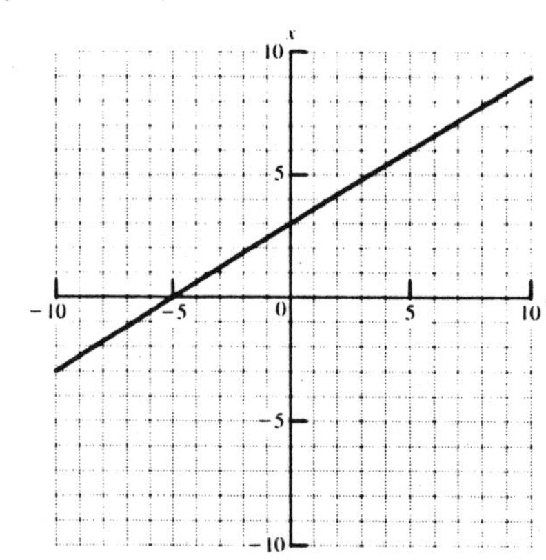

4.

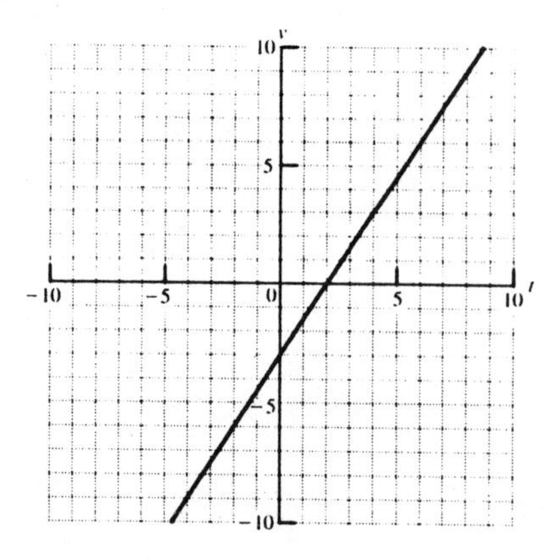

Find the slope and intercept for the lines below.

5.

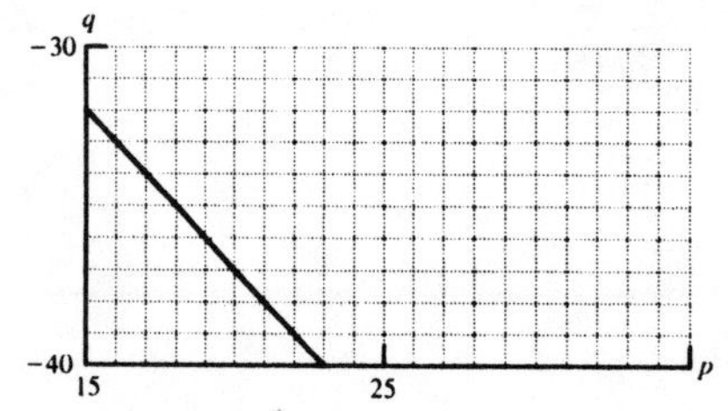

6.

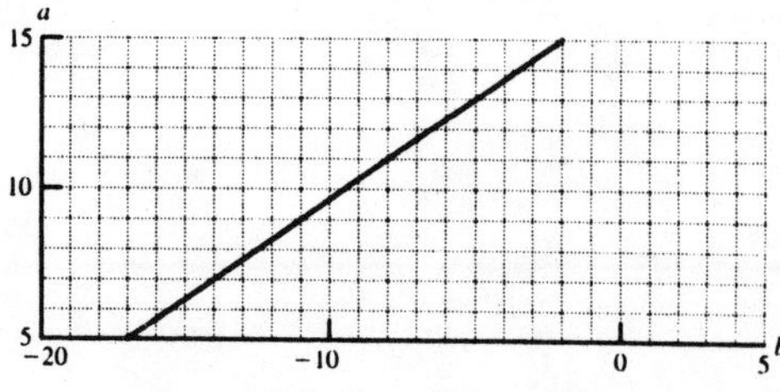

Draw the line corresponding to the equation given.

7. $y = 2x + 3$

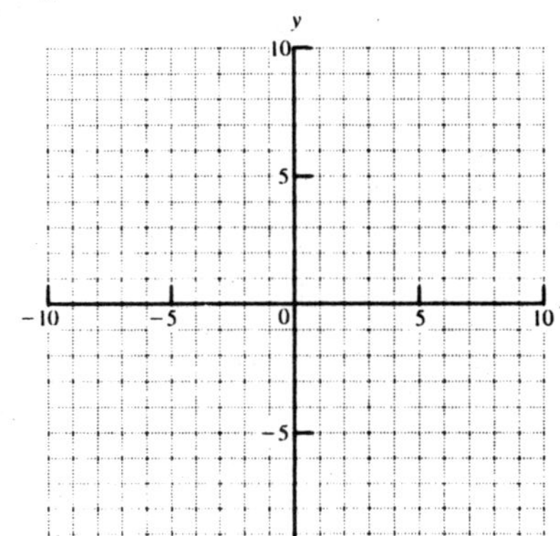

8. $p = 2t + 4$

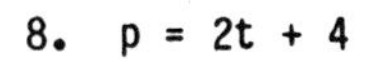

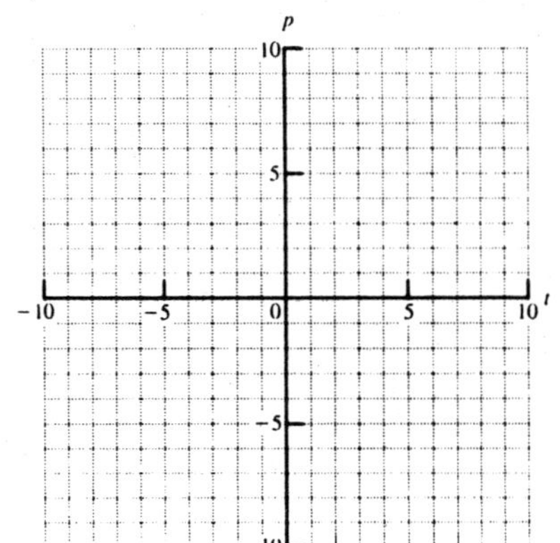

9. $v = -4t$

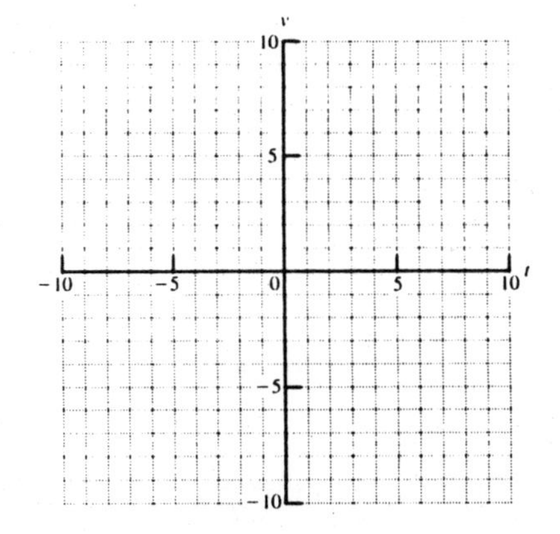

10. $a = 4$

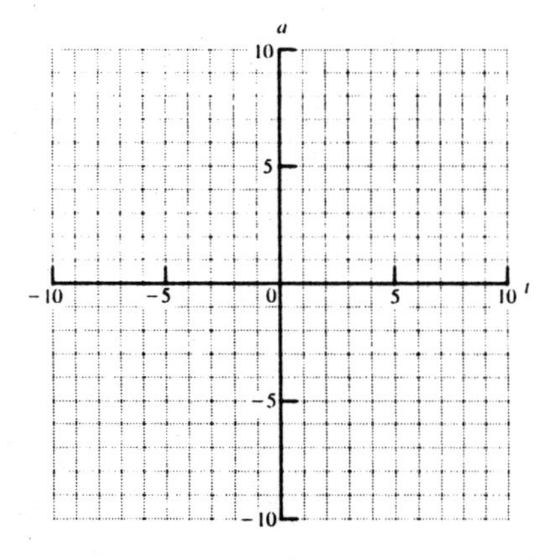

Find the slope and intercept for the lines below.

11.

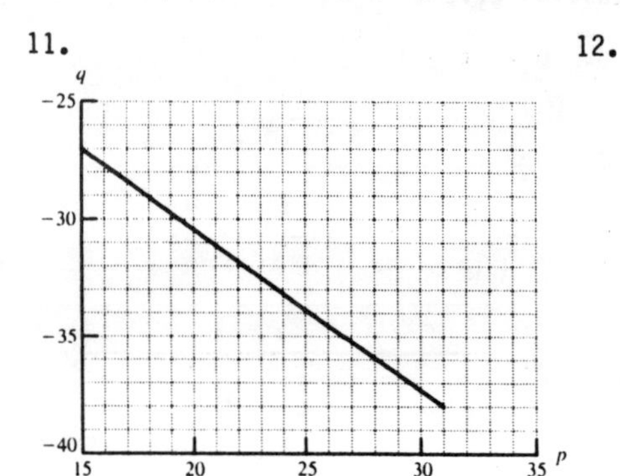

12.

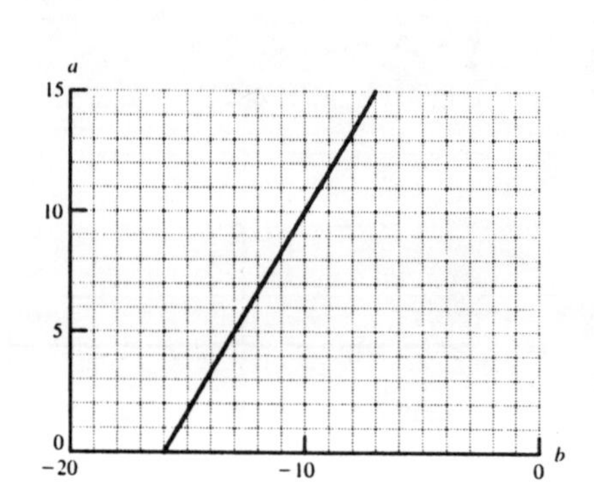

10.2 NONLINEAR GRAPHS

While the linear relationships between variables are important, most variables have nonlinear relationships. If the relation is not linear, there is no longer any general parameter that can be labeled the slope. As will be seen, the concept of a slope at a particular point can be applied to any curve, but for the general curve the slope defined in this way changes from point to point along the curve.

For example, a sine curve is depicted in the following graph. This curve may be written as $y = A \sin (\omega t + \delta)$ where A, ω, and δ are all constants (ω and δ are the Greek letters omega and delta, frequently used as symbols when discussing angular and rotational motion). Trigonometry is the topic of the next chapter, and in that chapter the sine function will be discussed in detail. In case you are curious, for the curve illustrated, $A = 10$, $\omega = \pi/10$, and $\delta = 37°$.

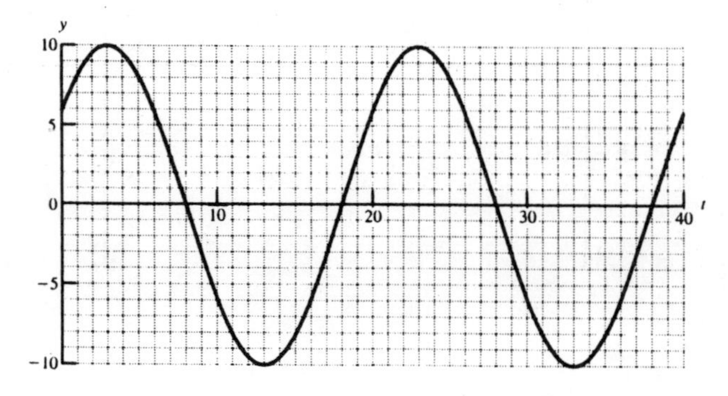

The question "What is the slope of this curve?" has no answer. "The" slope as a parameter for the curve does not exist. However, if the question is "What is the slope of the curve at t = 3?," the correct answer is zero.

It may look like there is a paradox here. In one sentence the statement appears that there is no slope value for the curve, and in the next sentence the slope is given a numerical value. If you didn't notice the difference in the two questions about slope, go back and read the preceding paragraph again.

In the second question there is a qualifier "...at t = 3?." The two questions are not the same. While the curve does not have a unique "slope," it does have something that can be called a slope at any point on the curve. This local slope is the rate of change of vertical variable to rate of change of horizontal variable over a very small range right at the point in question.

At first glance, it may seem a contradiction in terms to speak of a change in something at a point. However, consider what curves represent in the context of physical phenomena. If any phenomenon that actually occurs is examined in small enough detail, the curve will become linear over a reasonable distance of the graph. What this amounts to is expanding the horizontal and vertical scales to a greater and greater degree, making the increments on that scale represent smaller and smaller amounts. Eventually the representation on a normal size graph (for example, to fit the page of a book) will be a straight line. On such a graph, the slope is now defined (by y = mx + b), and determined by change in vertical variable divided by change in horizontal variable between two points on the line (which is actually a small segment of the more complex curve). The slope is the slope of the curve over the expanded range, and may be used as the slope at any point on the expanded range.

It may be bothersome that a nonlinear curve is turned into a linear curve by the witchcraft of expanding a scale. Perhaps you have seen very high speed photography where an event is photographed at 1000 frames a second and played back at 30 frames per second. In the process the time scale has been expanded 33 times and events that happen so close together that they normally cannot be separated become observable as separate. If a bullet passing through a soap bubble is observed with this technique, the bullet can be followed through the process of penetrating first one and then the other edge of the bubble, followed by the collapse of the bubble. Every detail is observable. On an expanded time scale, an event that happened in 1/30 of a second would be a discontinuity on a graph where one division of the scale represented one second. On a scale expanded to one division equal to 1/1000 of a second, the same event requiring 1/30 second would be spread out over 33 time scale divisions. The discontinuity that was crowded into a line that appeared to be vertical (meaning the entire event took place in a time less than the width of the trace defining the curve) now becomes a smooth curve. Consider the curves below. The time scale is expanded by 1000 around t = 0.

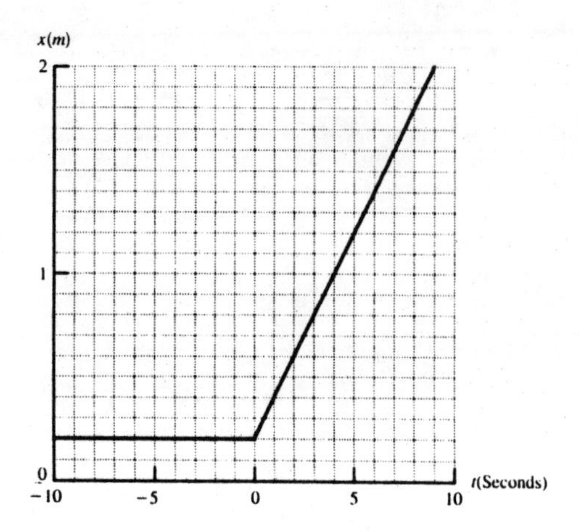

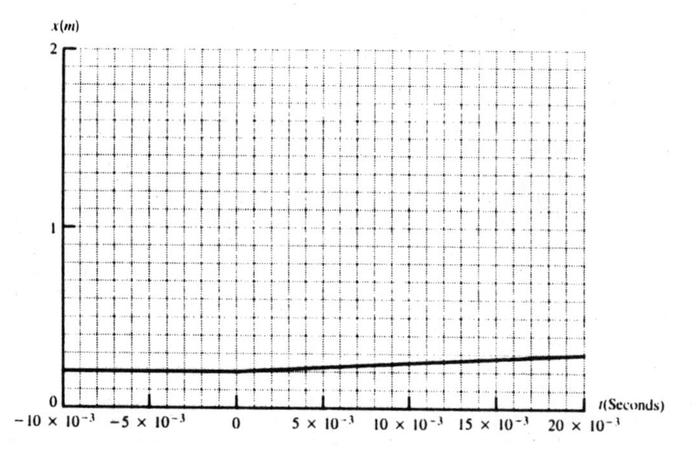

Of course, there is nothing magic to 1/1000 of a second, and the time scale might be expanded even more. If the time axis is expanded ten more times, ten units on the graph will contain everything that was originally contained in one, and the curve becomes even more linear.

As the time scale expands, the entire display of the graph represents less and less time. To the original scale, where each unit represented one second, the entire expanded display eventually will represent only an "instant." In this sense, the instant is one very small part on the time axis of the original graph.

We use concept of an "instant" all the time in our everyday life. If you leave home to go to a movie, you would be comfortable giving a specific time for the departure. Yet on a small enough time scale, the leaving process takes place over several units of the small time increments. On a small enough scale, the process may be described as a series of linear events, each having a definite slope over very small time periods.

The slope over the "instant", taken from the expanded display, is the slope of the tangent line to the curve at the time value on the original graph. In other words, if we desire the slope at any point on our time curve, we take the slope of the tangent line at the time in question. At some later time, the slope will change because the slope of the tangent line has changed.

Time is a very important parameter to physics, and when the slope of a curve of something plotted against time is found at a specific time this slope is called an instantaneous value of one kind or another. For example, the instantaneous velocity is the slope of the position-time curve (time being plotted on the horizontal axis). This is the value of the velocity at that instant.

Of course, all curves are not time curves. The process is general, and the slope of any curve at a given point is the slope of the tangent line at that point.

Rule: The slope of a general curve at a specific point on the curve is the slope of the tangent line to the curve at that point.

You have probably had occasion to observe that the slope at a given value of the coordinate is the slope of the tangent line. Imagine a curved highway. As a car drives this highway at night, the headlights of the car always point directly in the direction the car is moving. If the highway is drawn on a graph, with the axes representing directions such as north and east, the slope means change in one dimension divided by the change in the other, not a time rate of change or a vertical rise over a horizontal run. The tangent line to the curve at a point will parallel the slope of the curve at that point, and in turn the direction of the highway at the corresponding point on the highway. The lights might point out into the woods at a given point as the car rounds a curve. At the moment (the instant) the lights shine in the direction of travel, not to the point the car will eventually reach.

A straight line is a special case of a general curve, and the tangent line to a straight line is also the original line. For any linear relationship, the value of the slope at a given point is also the value of the slope over any extended displacement. If one axis is time, the instantaneous slope (of whatever the property is being determined) is the same as the slope of this same property over any time period.

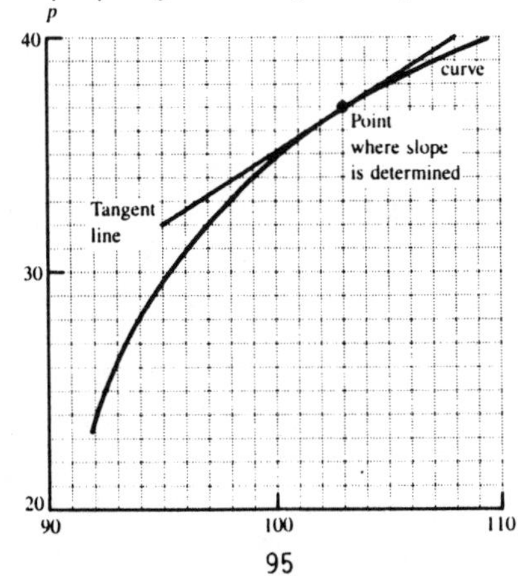

For a curve other than a straight line, drawing the tangent line is a judgment and subject to considerable error. The tangent is drawn by taking a straight edge and adjusting it until the angles between the straight edge and the curve look the same on the two sides. Being a judgment, the line may also be off to some degree and the slope correspondingly be in error. Numbers taken from graphs are always subject to considerable error.

When a curve is such as to change a large amount in one dimension for a small change in the other, the slope of the tangent line is especially subject to error. If the curve has discontinuities (that is, points or sharp turns), this means the scales are too large to determine anything at all about what happens during the period represented on the graph for the discontinuity. In such a case, if it is desired to know more about what happens over the region of the discontinuity, more information must be supplied than is represented on the graph.

As an example of how the slope of a general curve may be determined, consider the sine function below. The second curve is a reproduction of the first, with tangent lines at t = 4, t = 12, and t = 19.

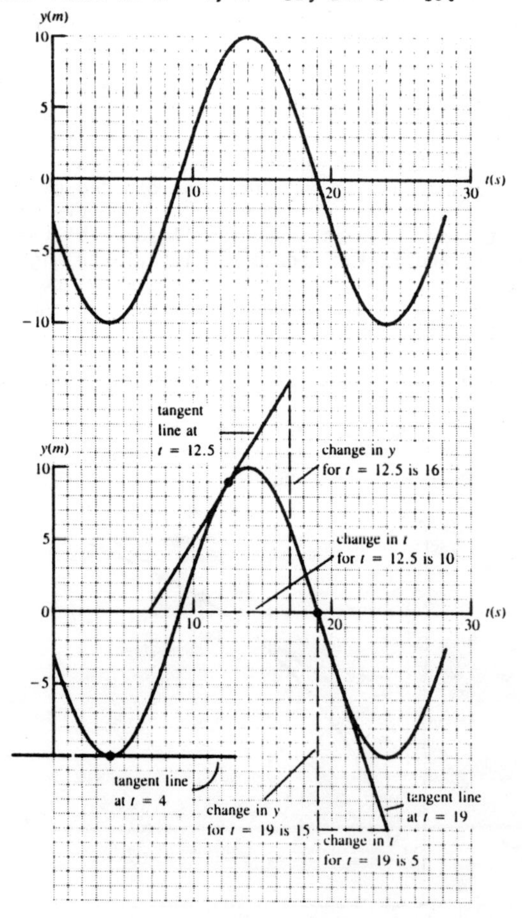

At $t = 4$, the tangent line is horizontal, corresponding to a slope of zero. Even though the value of the curve is changing rapidly at this point, it is relatively easy to see that the minimum occurs at $t = 4$, and for a minimum the slope must be zero right at this point.

At $t = 12.5$ the slope is not so exact. The line drawn is a reasonable line for a tangent line, and it is extended to make the respective changes in vertical and horizontal variable readable. For this line, the tangent line changes by +16 in y for a change of +11 in time, and the slope is 1.5 m/s.

At $t = 19$, the slope of the tangent line is given by a change of -15 in y for a change of +5 in time, and the slope is -3.0 m/s.

Now, as to accuracy: The equation for the curve illustrated is

$$y = 10 \sin (\pi t/10 + 1.1\pi),$$

or, expressing the angle in degrees,

$$y = 10 \sin (18°t + 198°).$$

The slope may be determined accurately using the calculus, and the actual values are

slope = 0 at $t = 4$ (compared with slope = 0 from graph),

slope = 1.6 at $t = 12.5$ (compared with slope = 1.5 from graph),

slope = -3.14 at $t = 19$ (compared with slope = -3.0 from graph).

As you can see, the values obtained from the graph are off in the second digit except at $t = 4$, a point of symmetry. If different people determine the slope at a point from a curve, there may be considerable difference between readings in the second digit. It is very difficult to draw the exact tangent line when the curve is not linear.

Practice Problems

Refer to the curve below for the practice problems.

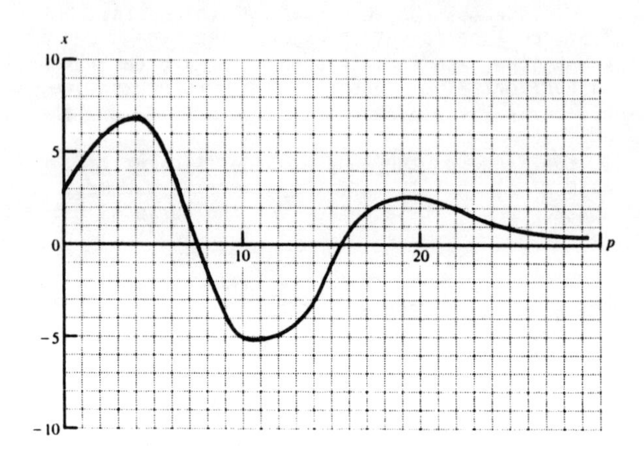

13. Find the slope of the curve at p = 4.

14. Find the slope of the curve at p = 13.

15. Find the slope of the curve at p = 15.3.

16. Find the slope of the curve at p = 22.7.

Answers to Practice Problems

1. $5y = -6x + 20$

2. $y = -5$

3. $5x = 3t + 15$

4. $2v = 3t - 6$

5. slope = -1
   intercept = -17

6. slope = 2/3
   intercept = 16.3

7. $y = 2x + 3$

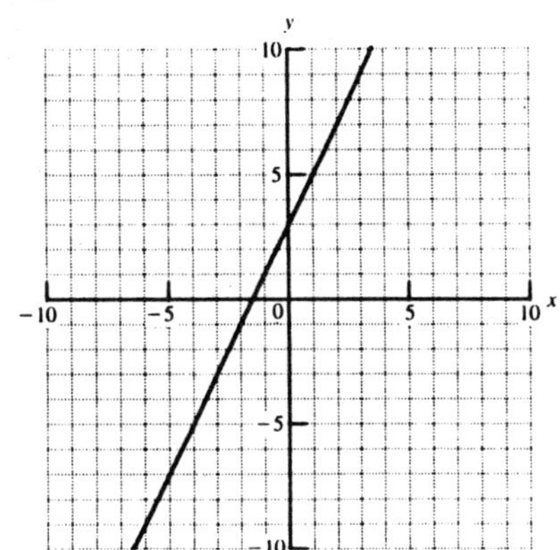

8. $p = 2t + 4$

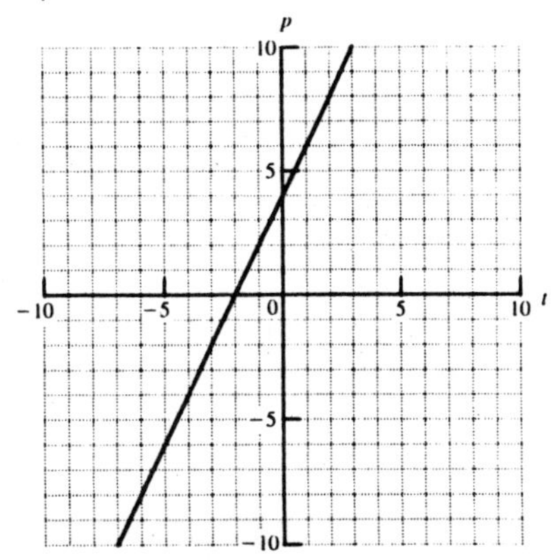

9. $v = -4t$

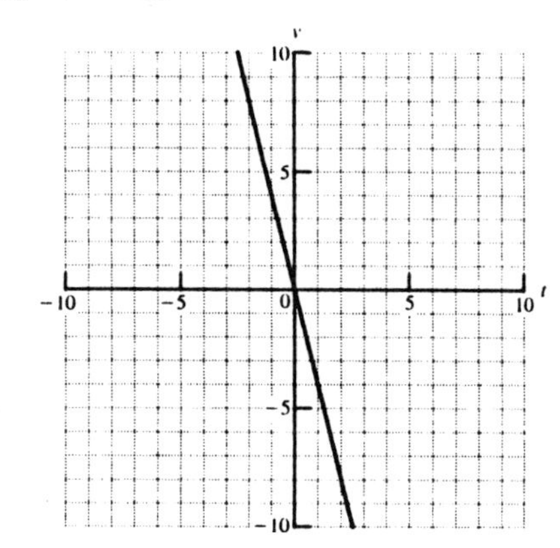

10. $a = 4$

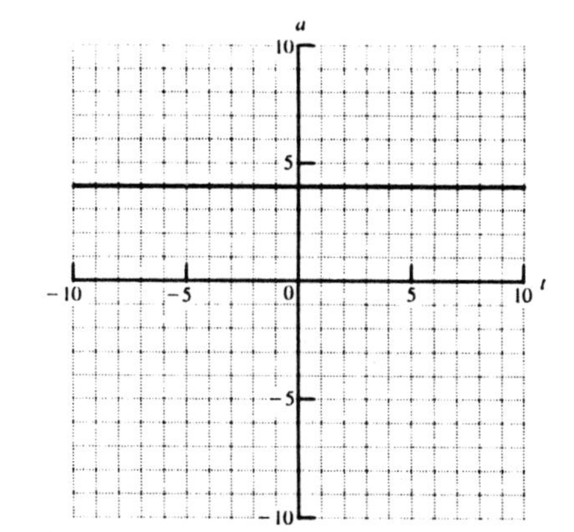

11. slope = -0.7
intercept = -17

12. slope = 1.7
intercept = 27

13. zero

14. 0.9

15. 2

16. -0.5

Exercises

Write the equation for each of the curves illustrated.

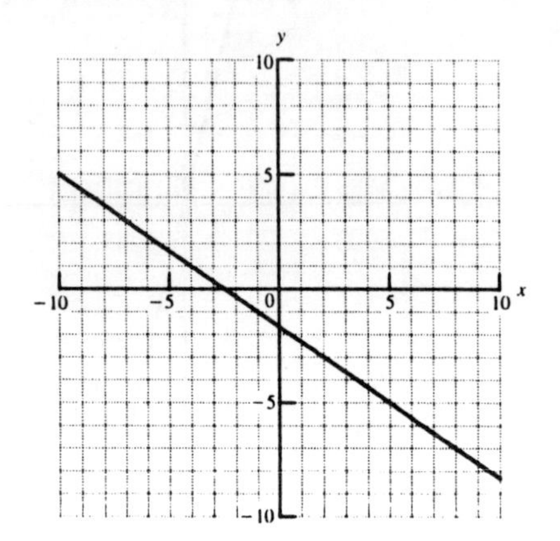

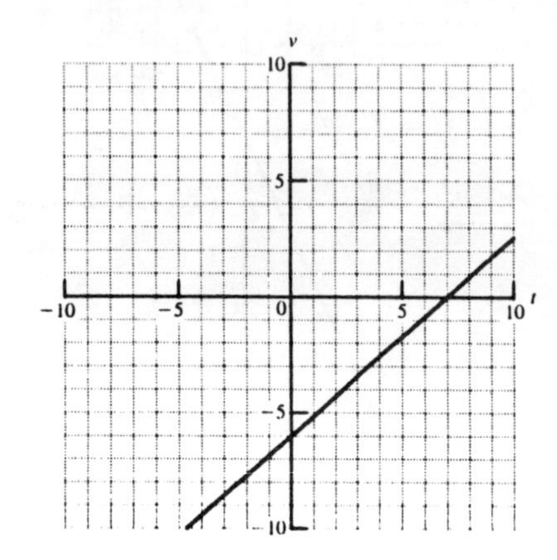

1. ____________ 2. ________________

Plot the curve corresponding to the equation given.

3. $x = -4 + 3t$

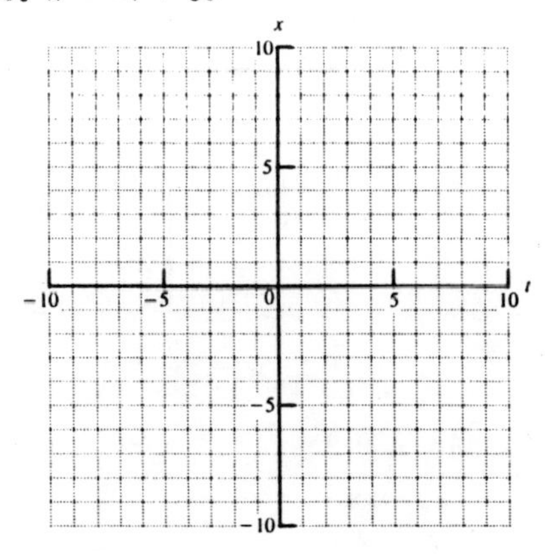

4. $p = -2q - 5$

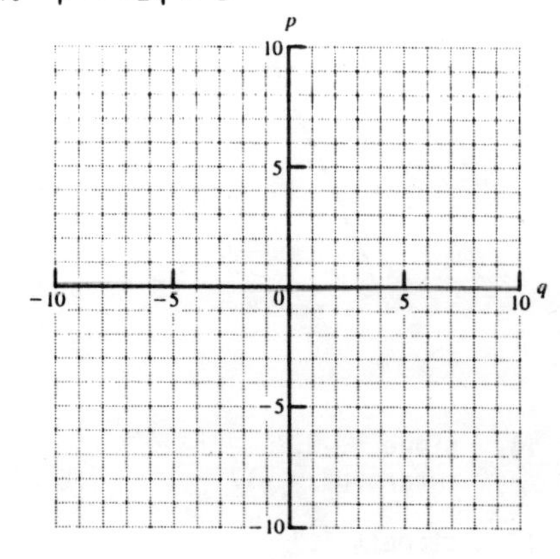

5. $a + 2 = b - 5$

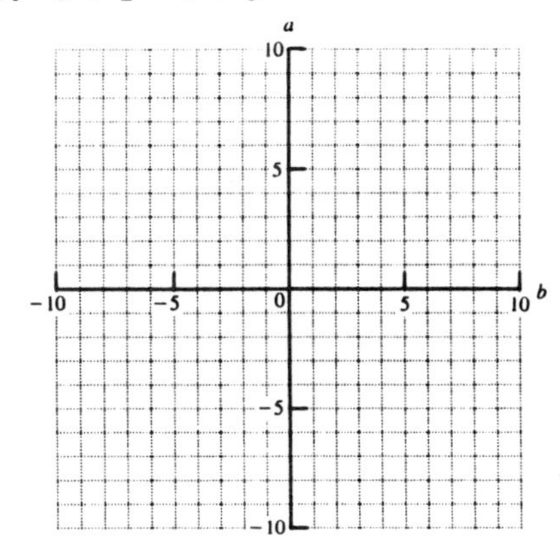

6. $3x - 4 = 2(y + 2)$

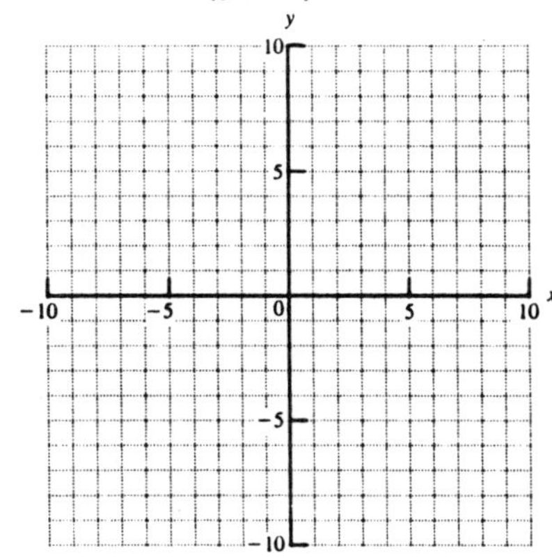

Find the slope and intercept of the lines below.

7.

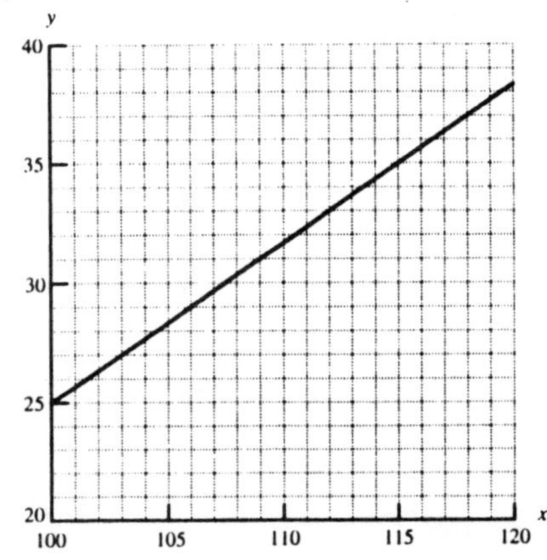

8.

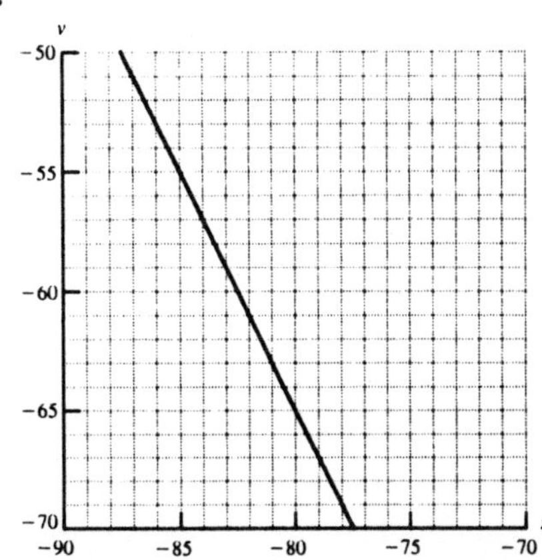

Refer to the curve below for questions 9 through 12.

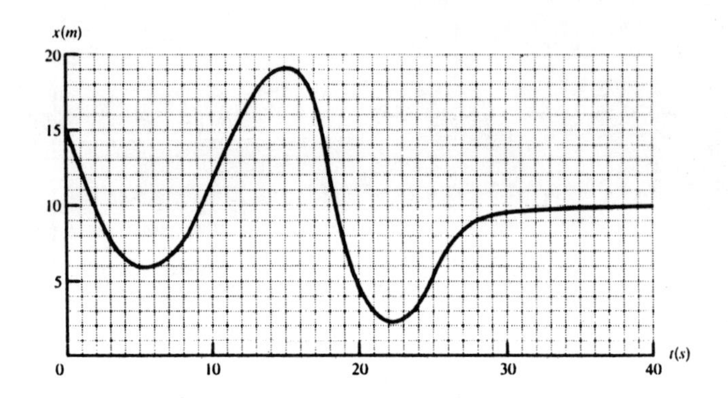

9. At what times is the slope zero?
10. What is the slope at t = 10?
11. What is the slope at t = 25?
12. What is the slope at t = 2?

11

Trigonometric Functions

Geometrical relationships occur over and over in the study of physics. While the rules of algebra are followed when handling equations, the rules of trigonometry are followed when handling geometric relationships. Any physics course covering material at any significant depth will expect you to be able to work with trigonometric relationships.

11.1 PYTHAGOREAN THEOREM

Trigonometry is based on right triangles. A right triangle is any triangle having one angle of 90°. The 90° angle is called the right angle, and the side opposite the right angle is called the hypotenuse. A special relationship exists between the sides of a right triangle. This relationship is called the Pythagorean Theorem.

> Rule (the Pythagorean Theorem): In a right triangle having sides a, b, and c units long, the sides are related to each other according to
>
> $a^2 + b^2 = c^2$, where c is the side opposite the right angle.

The Pythagorean theorem may be used to find the third side of a right triangle, given the other two sides. For example, the triangle illustrated on the top of the next page has sides of length 3 and 4. The hypotenuse is given by $\sqrt{3^2 + 4^2} = 5$.

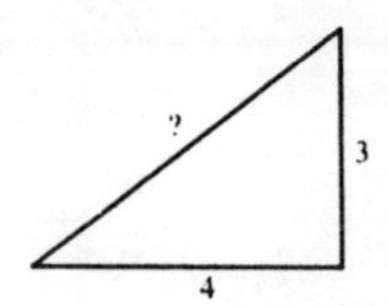

Of course, if any two sides are given, the third side may be found using the Pythagorean Theorem. In the triangle illustrated below the hypotenuse is 2 units long and one of the sides is 1 unit long. The unknown side is $\sqrt{2^2 - 1^2} = \sqrt{3}$ units.

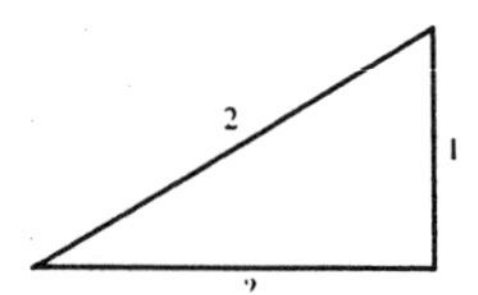

Practice Problems

Find the unknown side indicated by ?.

1.

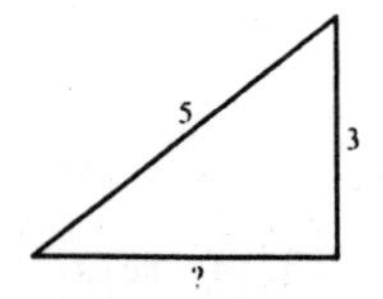

2.

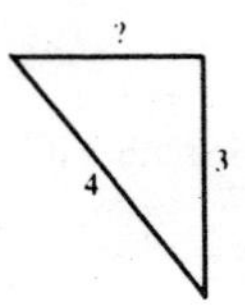

3.

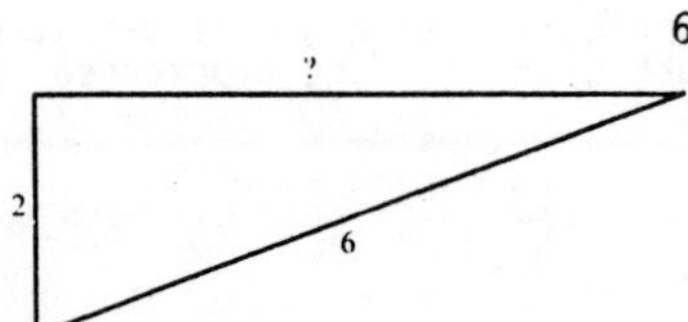

4.

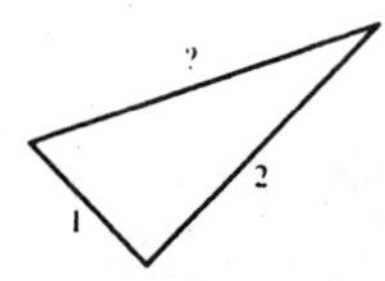

5.

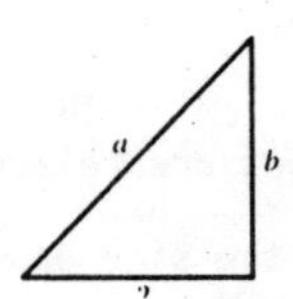

6.

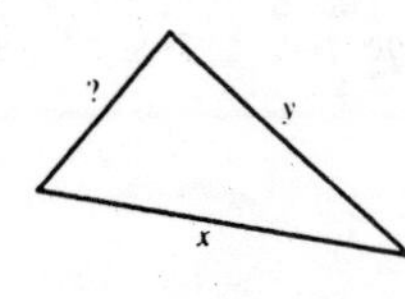

7.

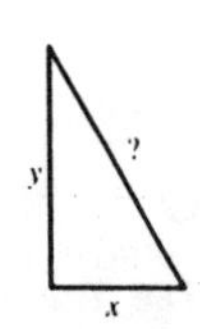

8.

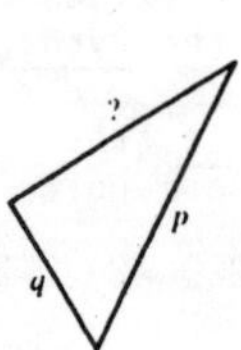

11.2 RADIAN MEASURE

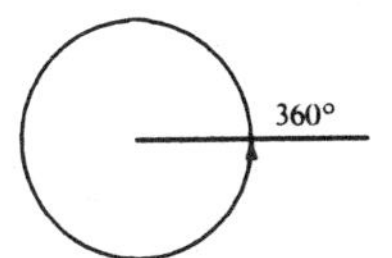

Angles are usually measured in degrees or radians. The degree is defined by dividing a full circle into 360 equal angular parts. One degree is 1/360th of a full circle. The degree is dimensionless. The symbol for "degree" is a small superscripted circle. For example, 30 degrees is written 30°.

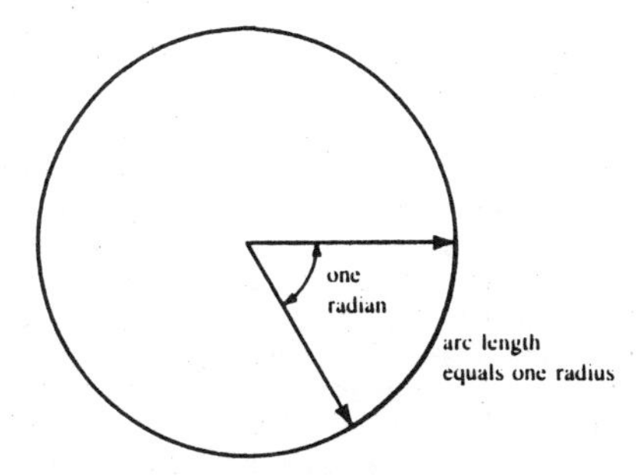

The radian is based on the ratio of the length of an arc on the circumference of a circle to the radius of the circle. When the arc length is equal to the radius, the angle subtended is 1 radian. Since the circumference of a circle is 2 times pi times the radius ($2\pi r$), a circle contains 2π radians. Radians are often expressed as a multiple of pi, although it is perfectly acceptable to use simple numbers.

Rule: $360° = 2\pi$ radians.

Degrees may be converted to radians and radians to degrees by application of the above rule. Simply multiply degrees by $2\pi/360$ to obtain radians and multiply radians by $360°/2\pi$ to obtain degrees.

$30° = (30)(2\pi/360) = \pi/6$ radians.

$45° = (45)(2\pi/360) = \pi/4$ radians.

$\pi/2$ radians $= (\pi/2)(360/2\pi) = 90°$.

Practice Problems

Convert to degrees:

9. $2\pi/6$ radians

10. $4\pi/5$ radians

11. 1 radian

12. $\pi/8$ radians

Convert to radians.

13. 1°

14. 30°

15. 270°

16. 15°

11.3 TRIGONOMETRIC FUNCTIONS

In this workbook, we will discuss only the three fundamental trigonometric functions: the sine, the cosine, and the tangent. The reciprocals of these three functions are of course functions as well, and are given names: 1/sine = cosecant, 1/cosine = secant, and 1/tangent = cotangent.

The trigonometric functions are based on geometric relationships in a set of mutually perpendicular coordinates, almost invariably labeled x and y. In such a system, four quadrants are present and an angle may lie in any of these quadrants. The trigonometric functions of an angle are all defined in terms of the angle between the positive x-axis and a line from the origin to a point having coordinates x and y. By convention, angles are measured positive counterclockwise from the x-axis. The x and y axes divide the plane into the four regions called quadrants. Lines in the four quadrants and the angle with positive x-axis (labeled θ) are illustrated in the following four graphs.

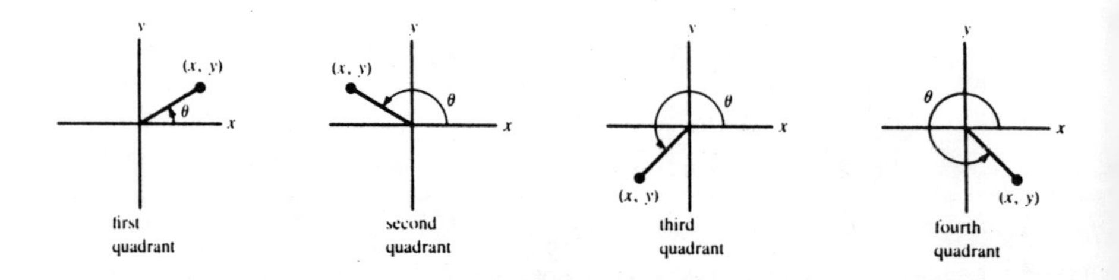

The trigonometric functions may be defined by a triangle in the first quadrant as shown in the illustration below. The line from the origin to the point P (with coordinates x and y) makes an angle θ with the positive direction along the x-axis. The length of the line is labeled r and is, by the Pythagorean Theorem, equal to $\sqrt{x^2 + y^2}$.

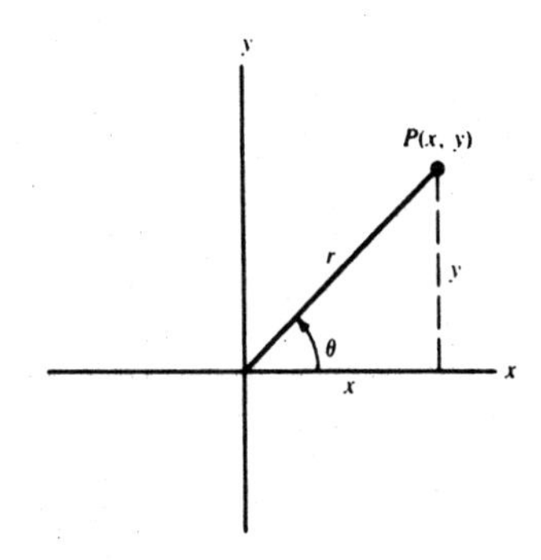

The trigonometric functions are defined as follows:

Sine θ (abbreviated sin θ) = y/r

Cosine θ (abbreviated cos θ) = x/r

Tangent θ (abbreviated tan θ) = y/x

Notice that the tangent may be expressed in terms of the sine and cosine:

$$\tan\theta = y/x = (y/r)/(x/r) = (\sin\theta)/(\cos\theta).$$

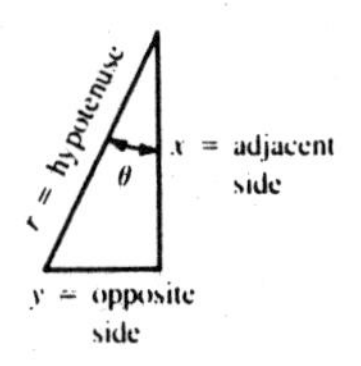

The functions may be defined in terms of the ratios of sides and hypotenuse of a right triangle of any orientation. For the triangle illustrated above, the trigonometric functions of the angle θ are

$$\sin\theta = \frac{\text{opposite side}}{\text{hypotenuse}} = y/r,$$

$$\cos\theta = \frac{\text{adjacent side}}{\text{hypotenuse}} = x/r,$$

$$\tan\theta = \frac{\text{opposite side}}{\text{adjacent side}} = y/x.$$

Practice Problems

Find the trigonometric functions (sine, cosine, and tangent) for the angles indicated in the triangles. All of the triangles are right triangles.

17.

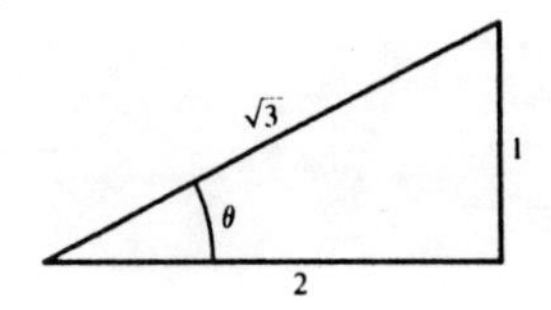

$\sin\theta =$ ________

$\cos\theta =$ ________

$\tan\theta =$ ________

18.

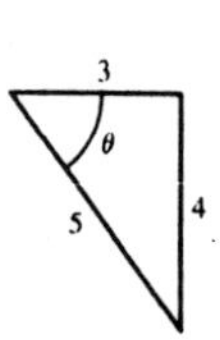

$\sin\theta =$ ________

$\cos\theta =$ ________

$\tan\theta =$ ________

19.

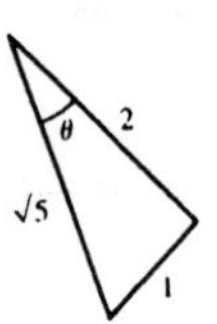

$\sin\theta =$ ________

$\cos\theta =$ ________

$\tan\theta =$ ________

20.

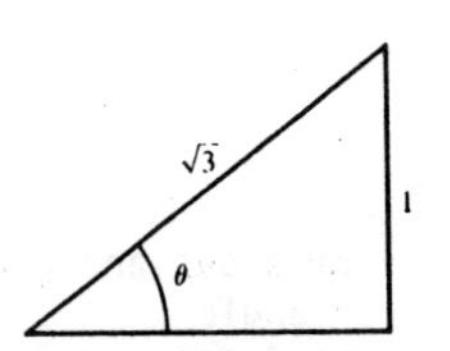

$\sin\theta =$ ________

$\cos\theta =$ ________

$\tan\theta =$ ________

21.

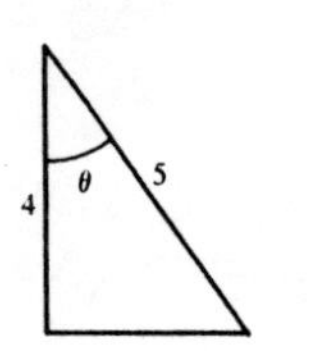

$\sin\theta =$ ________

$\cos\theta =$ ________

$\tan\theta =$ ________

22.

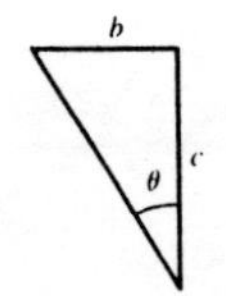

$\sin\theta =$ ________

$\cos\theta =$ ________

$\tan\theta =$ ________

23

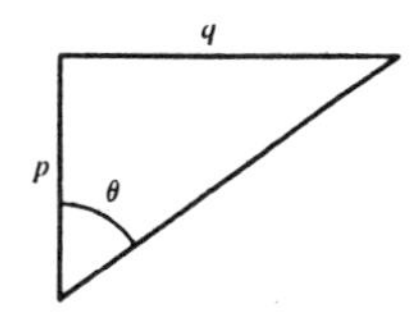

sin θ = ________

cos θ = ________

tan θ = ________

24.

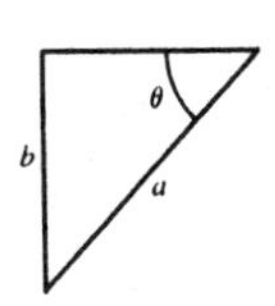

sin θ = ________

cos θ = ________

tan θ = ________

11.4 ANGLES LARGER THAN 90°

It is impossible to construct a right triangle with an angle larger than 90°, so trigonometric functions for angles in the second, third, and fourth quadrants cannot be defined in terms of the sides of a triangle. In these quadrants, the defining relations for the functions are based on the definitions given in section 11.3 with regard to x, y, and r. The distance r from the origin to the point P is always considered to be positive. x and y may be positive or negative depending on the quadrant.

In the second quadrant, x is negative and y is positive. Sin θ = y/r will therefore be positive. Cos θ = x/r is negative (because x is negative), and tan θ is negative.

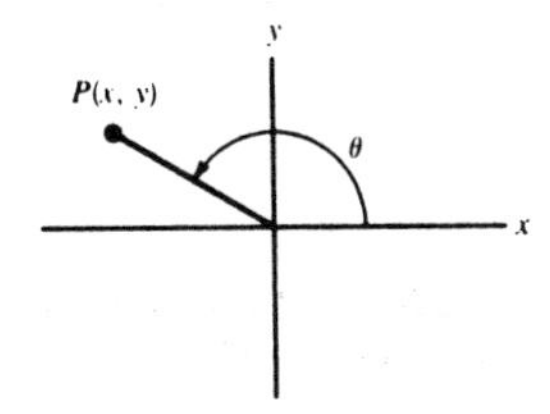

Two angles that add to 180° are said to be <u>complementary</u>, and for complementary angles the ratio $|x|/|y|$ is the same. This means that in the second quadrant, sin θ = sin (180° - θ). (For angle measure in radians sin θ = sin [π - θ].)

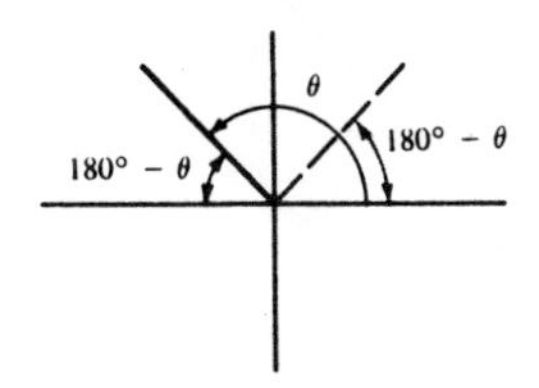

In the second quadrant:

$$\sin\theta = \sin(180^\circ - \theta) = \sin(\pi - \theta),$$

$$\cos\theta = -\cos(180^\circ - \theta) = -\cos(\pi - \theta),$$

$$\tan\theta = (\sin\theta)/(\cos\theta) = -\tan(180^\circ - \theta) = -\tan(\pi - \theta).$$

For example, $\sin 150^\circ = +\sin 30^\circ$; $\cos 150^\circ = -\cos 30^\circ$; $\tan 150^\circ = -\tan 30^\circ$.

The sine and cosine functions are illustrated below with the second quadrant angular region emphasized. Notice that the sine function is symmetric about $\pi/2$ (90°). The cosine function is zero at 90°, and the values of the function in the second quadrant are antisymmetric about 90° (meaning the values are equal in magnitude and opposite in sign for angles equidistant from 90°).

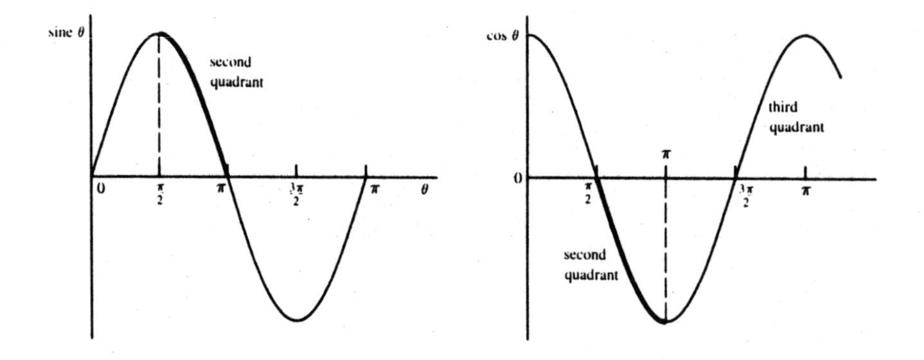

In the third quadrant, both x and y are negative. The sine and cosine are both negative in this quadrant, and the tangent is positive.

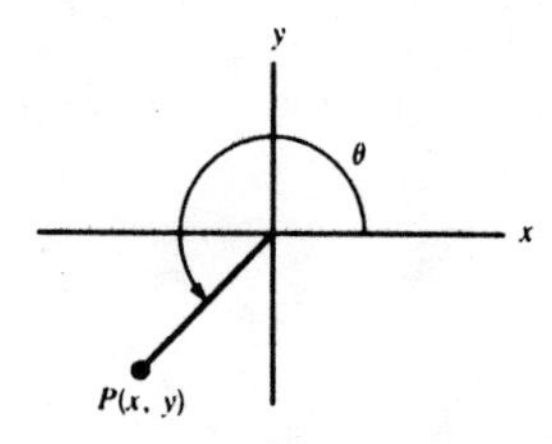

If the line is extended back into the first quadrant, it may be seen that the ratio of x to y for points on the line is the same for the original or the extended line.

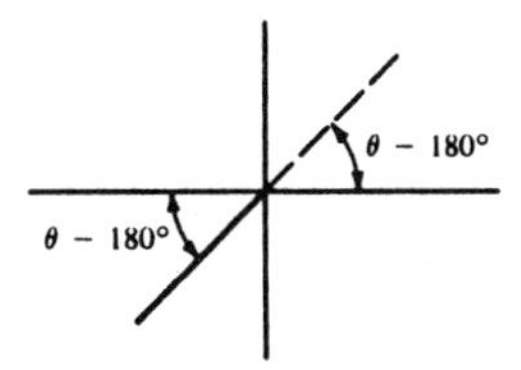

The trigonometric functions for an angle θ in the third quadrant are based on the functions of $(\theta - 180°)$. In the third quadrant:

$$\sin \theta = -\sin (\theta - 180°) = -\sin (\theta - \pi),$$

$$\cos \theta = -\cos (\theta - 180°) = -\cos (\theta - \pi),$$

$$\tan \theta = \tan (\theta - 180°) = \tan (\theta - \pi).$$

For example, $\sin 225° = \sin (5\pi/4) = -\sin 45° = -\sin (\pi/4)$. $\text{Cos } 225° = -\cos 45°$ and $\tan 225° = +\tan 45°$.

Referring again to the sine and cosine functions (illustrated in the following two graphs), it may be seen that in the third quadrant both sine and cosine functions reproduce the negative of the values for the functions in the first quadrant. The sine starts at zero at 180° and goes to -1 at 270° while the cosine starts at -1 at 180° and goes to zero at 270°. For any angle θ in this quadrant, $\sin \theta = -\sin (\theta - \pi)$ and $\cos \theta = -\cos (\theta - \pi)$.

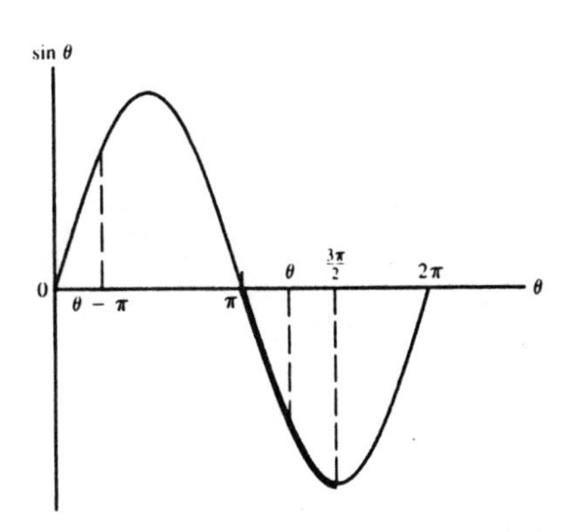

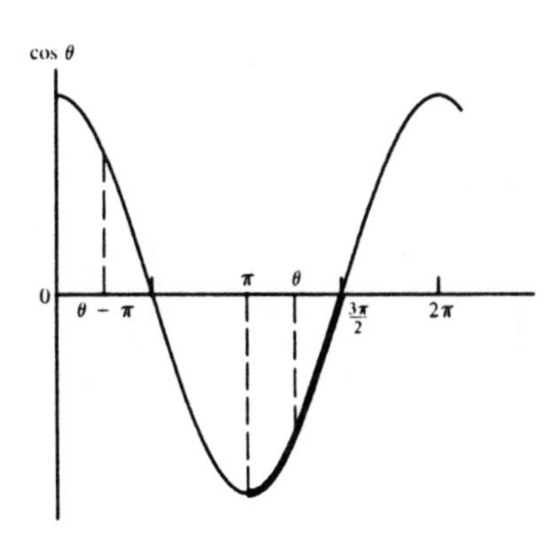

In the fourth quadrant, x is positive and y is negative. It then follows that the cosine is positive, the sine and the tangent are negative.

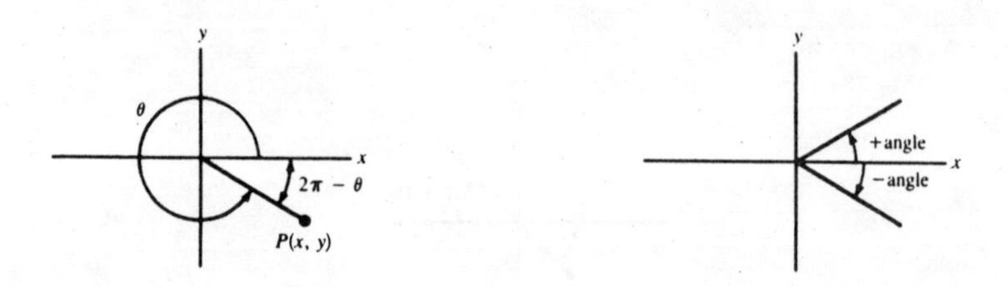

Also, since a full circle is 360° or 2π radians, the trigonometric functions of an angle are equal to the functions of the negative of 2π - the angle. In other words, the trigonometric functions of 330° are the same as the functions of -30°. For an angle θ in the fourth quadrant,

$$\sin \theta = -\sin (2\pi - \theta),$$

$$\cos \theta = \cos (2\pi - \theta),$$

$$\tan \theta = -\tan (2\pi - \theta).$$

For example, sin 300° = -sin 60°; cos 300° = +cos 60°; tan 300° = tan (5π/6) = -tan (π/3).

The sine and cosine curves are reproduced in the following graphs with the fourth quadrant emphasized. It may be seen that the sine is negative, and that the value of the sine for an angle θ is equal to the negative of the sine of the angle (2π - θ). The cosine is positive in the fourth quadrant, and it may be seen that the cosine of an angle θ is equal to the cosine of (2π - θ).

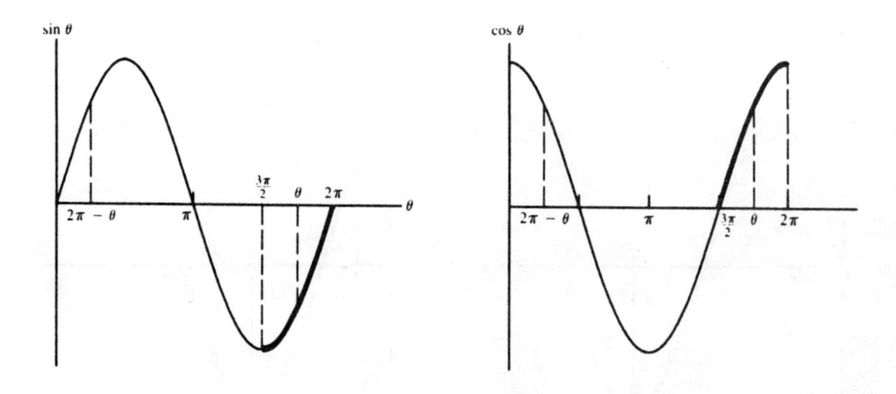

Trigonometric functions repeat on a cycle of 2π radians or 360°. This is because a circle repeats after one full revolution. This means that an angle 390° has the same trigonometric functions as 30°, 415° as 75°, 700° as 240°, and so on.

Negative angles are generated when the functions are extended in the negative direction. The sine function is antisymmetric about the zero in angle and the cosine function is symmetric about the zero in angle. Thus, as has been discussed, $\sin(-\theta) = -\sin\theta$ and $\cos(-\theta) = \cos\theta$. The sine and cosine functions are reproduced below for several cycles.

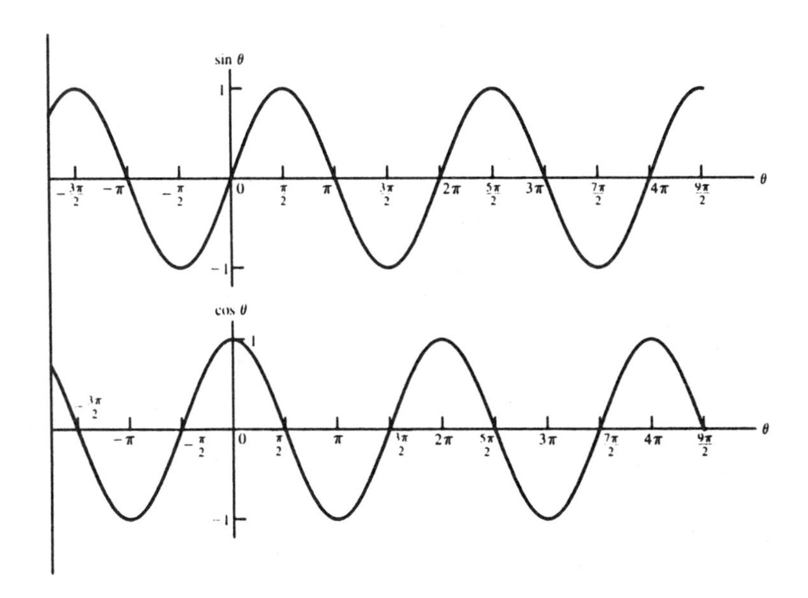

Physics texts and instructors tend to use positive and negative angles freely, and it will be useful to remember the rules for signs on the trigonometric functions when the angles are negative.

Rule: When the argument of an angle is negative, the trigonometric functions are related to functions of positive argument according to: $\sin(-\theta) = -\sin\theta$; $\cos(-\theta) = \cos\theta$; $\tan(-\theta) = -\tan\theta$.

The tangent function has not been treated in depth. This is because the tangent is not as widely used as the sine and cosine. For those times when you will need to use the tangent function as more than a division of sine and cosine, the function is reproduced on the next page.

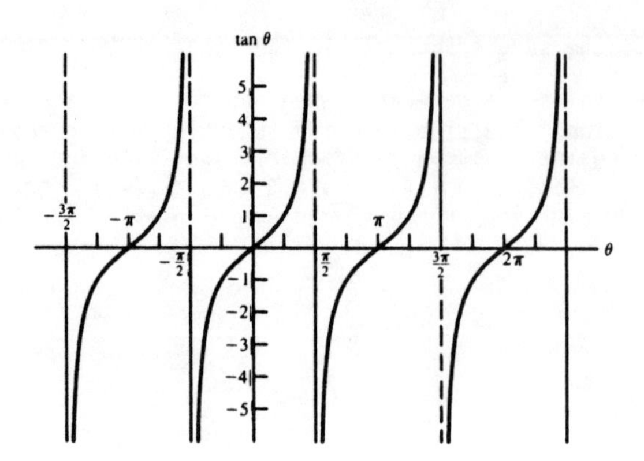

Notice that at $\pi/2$ and $3\pi/2$ (90° and 270°), the value of the tangent goes to + or - infinity. At these angles the cosine is zero. The sign of the tangent function in any quadrant is determined by the signs of the sine and cosine according to $\tan\theta = (\sin\theta)/(\cos\theta)$.

Practice Problems

For the angles below, express the trigonometric functions in terms of an angle in the first quadrant.

25.

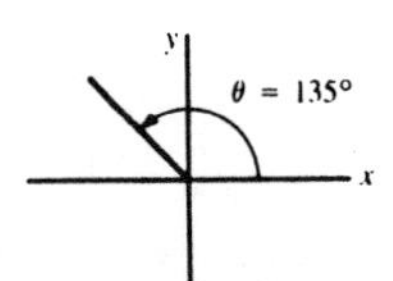

$\cos\theta$ = ____________

$\sin\theta$ = ____________

$\tan\theta$ = ____________

26.

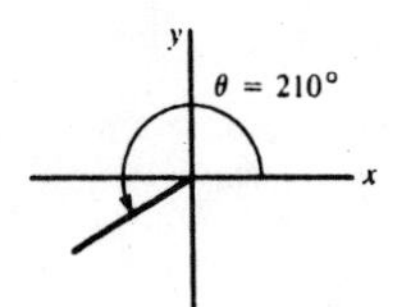

$\sin\theta$ = ____________

$\cos\theta$ = ____________

$\tan\theta$ = ____________

27.

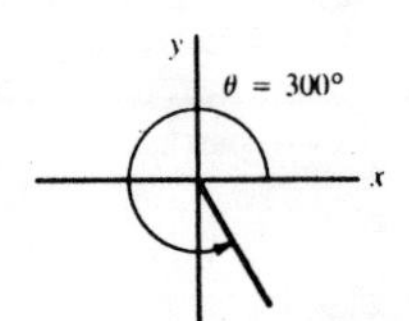

$\tan\theta$ = ____________

$\sin\theta$ = ____________

$\cos\theta$ = ____________

28.

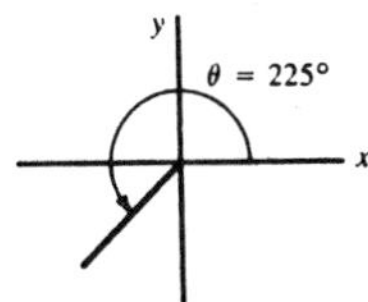

cos θ = ________

tan θ = ________

sin θ = ________

Express the angle between the line from the origin to the point indicated and the x-axis in terms of the coordinates of the point.

29.

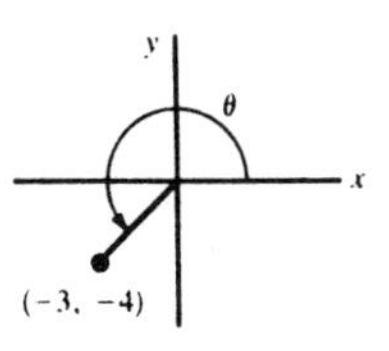

sin θ = ________

cos θ = ________

tan θ = ________

30.

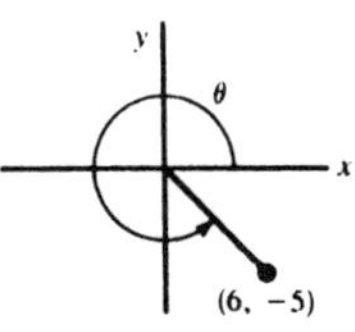

sin θ = ________

cos θ = ________

tan θ = ________

Express in terms of the function of an angle in the first quadrant.

31. sin (-30°) = ________

32. cos (-210°) = ________

33. tan (-300°) = ________

34. cos (5π/2) = ________

35. sin 750° = ________

36. cos (25π/6) = ________

11.5 MISCELLANEOUS TRIGONOMETRIC RELATIONS

The Pythagorean Theorem gives a relationship between the sides and hypotenuse of a right triangle. The trigonometric functions also relate the sides and hypotenuse of a right triangle. It should not come as a surprise that there are relationships between the various trigonometric functions beyond the definition of the tangent. In the formal study of trigonometry, many interrelationships between the various trigonometric functions are used. While all of these are important and most are eventually used if physics is studied at an advanced level, only a few are common to the introductory physics course. These most common relations are discussed next.

In a right triangle, the nonright angles have a special relationship that may be seen from the following triangle. For the angles α and β, the side adjacent to α is opposite to β, and $\sin \alpha = a/b = \cos \beta$.

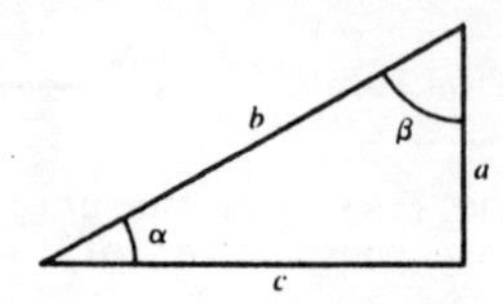

Since the sum of the angles of a triangle is 180°, in a right triangle the two angles other than the right triangle must add to 90°. When two angles add to 90°, they are said to be <u>complementary</u> and the sine of one is equal to the cosine of the other.

<u>Rule</u>: The sine of an angle less than 90° is equal to the cosine of the complementary angle.

<u>Practice Problems</u>

37. sin 30° = cos __________

38. sin 45° = cos __________

39. sin 70° = cos __________

40. cos 15° = sin __________

41. cos 88° = sin __________

42. cos 57° = sin __________

Another very useful relation between sine and cosine comes directly from the Pythagorean Theorem. In the following triangle, $r^2 = x^2 + y^2$, and $x^2/r^2 + y^2/r^2 = 1$.

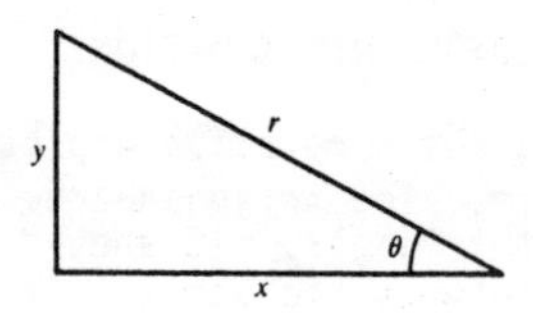

As the trigonometric functions are $\sin \theta = y/r$, $\cos \theta = x/r$, the preceding equation is equivalent to

$$(\cos \theta)^2 + (\sin \theta)^2 = 1.$$

Rule: For any angle θ, $(\sin\theta)^2 + (\cos\theta)^2 = 1$.

Very often the squares of the trigonometric functions are written differently from the manner above. $(\sin\theta)^2$ may equally well be written as $\sin^2\theta$. Similarly, $(\cos\theta)^2 = \cos^2\theta$ and $(\tan\theta)^2 = \tan^2\theta$. An alternative way of writing the equation $(\sin\theta)^2 + (\cos\theta)^2 = 1$ is

$$\sin^2\theta + \cos^2\theta = 1.$$

Dividing this equation by $\cos^2\theta$ gives

$$\frac{\sin^2\theta}{\cos^2\theta} + \frac{\cos^2\theta}{\cos^2\theta} = \frac{1}{\cos^2\theta},$$

or

$$\tan^2\theta + 1 = \frac{1}{\cos^2\theta}.$$

Division of $\sin^2\theta + \cos^2\theta = 1$ by $\sin^2\theta$ gives

$$1 + \frac{\cos^2\theta}{\sin^2\theta} = \frac{1}{\sin^2\theta}$$

$$1 + \frac{1}{\tan^2\theta} = \frac{1}{\sin^2\theta}.$$

Practice Problems

43. Given that $\sin 30° = 0.5$, find

$\cos 30° =$ __________ and $\tan 30° =$ _______________.

44. If $\cos\beta = 0.3$, $\sin\beta =$ _________ and $\tan\beta =$ ____________.

45. If $\tan\lambda = 2.0$, $\sin\lambda =$ ___________ and $\cos\lambda =$ ____________.

We have seen that for any two complementary angles α and β, $\sin\alpha = \cos\beta$. However, the requirement 90° between the angles is satisfied by either of two conditions, $\beta = (90° - \alpha)$ or $\beta = (90° + \alpha)$. As the sign of the cosine function changes by quadrant differently from the sine, only one of the conditions is correct.

The sine and cosine functions are reproduced in the next graph, superimposed (with cosine slightly shifted as dashed curve) so that you may see the relationship between them. Notice that the cosine function is shifted forward in angle with respect to the sine, so that the cosine of any angle θ is equal to the sine of that angle plus 90°.

Rule: For any angle θ, $\cos \theta = \sin (\theta + 90°)$.

For example, the sine of 150° is equal to the cosine of 60°. The angle labeled θ in the drawing is at 150° on the sine curve and 60° on the cosine curve.

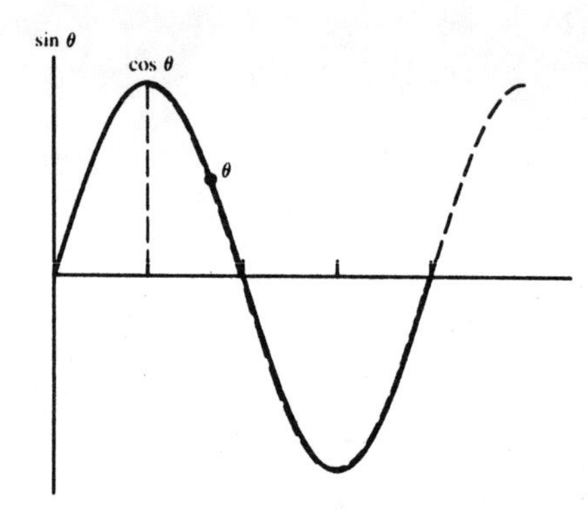

Of course, it may equally well be considered that the sine function is equal to the cosine function shifted backwards by 90°. By studying the figure, it should be apparent that either of the following relationships is valid.

$\sin (\theta + 90°) = \cos \theta$.

$\cos (\theta - 90°) = \sin \theta$.

Practice Problems

For the exercises below, make the direct conversion using $\sin (\theta + 90°) = \cos \theta$ and $\cos (\theta - 90°) = \sin \theta$.

46. sin 15° = cos ________

47. cos 315° = ________

48. sin 180° = cos ________

49. cos 150° = ________

11.6 INVERSE TRIGONOMETRIC FUNCTIONS

The trigonometric functions have a fixed value for a given angle. Conversely, a given value for a trigonometric function fixes the angle to one of two possibilities. These are called inverse functions.

The nomenclature is a little confusing for the inverse functions. The statement, "The angle whose sine is 0.70 is ...", is written "arcsin 0.70 = ..." or "$\sin^{-1}$ 0.70 = ...". The inverse function is identified by either the prefix "arc" or the exponent -1.

Inverse functions follow naturally from the values of the trigonometric functions. If sin 30° = 0.500, then arcsin (0.500) = 30°. If cos 45° = 0.707, then arccos 0.707 = 45°. If tan 60° = 1.73, then arctan 1.73 = 60°.

Trigonometric functions have the property that for any function, two angles have the same value. sin 30° = 0.500 and sin 150° = 0.500. cos 20° = 0.940 and cos 340° = 0.940. tan 35° = 0.700 and tan 215° = 0.700. As the signs change differently for the functions by quadrant, the different functions will have different angles with the same sine, cosine, or tangent. In order for you to know which angle is the correct angle for inverse functions, you must examine the problem on which you are working.

It is not the purpose of this workbook to explain use of calculators. Even so, most students of physics use these devices and the inverse trigonometric functions may be confusing. Many different brands of scientific calculators are available that have the trigonometric functions. These calculators will have a key labeled "INV" (for inverse). If you have such a calculator, punch the key sequence

. 5 [INV] [SIN] .

Your calculator should display 30 if in degree mode and 0.5236 if in radian mode. The angle whose sine is 0.5 is 30° or 0.5236 radians.

Calculators have the disadvantage thay they can display only one value. The default value for inverse trigonometric functions is the value of the angle in the first quadrant for positive arguments. For negative arguments, results are unpredictable. The angle may be negative (for example, arcsin (-.5) = -30°) or the value of the angle in the second quadrant. If you were asked for the angle between the x-axis and the line from the origin to the point (-4,-3), you could ask your calculator for this angle in several ways.

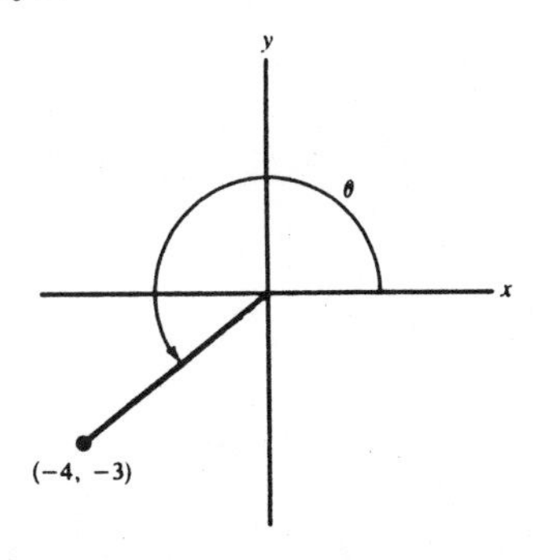

Arctan $\theta = (-3)/(-4)$ arcsin $\theta = (-3)/5$ or arccos $\theta = (-4)/5$. Your calculator will probably respond as follows (these are the response of a Texas Instruments TI 59 calculator):

$$\arctan 3/4 = 36.87°,$$

$$\arcsin (-3)/5 = -36.87°,$$

$$\arccos (-4)/5 = 143.13.$$

Of these, none are in the third quadrant, and none are correct as written. The correct angle is 216.87°, which is obtained from the above angles by examination.

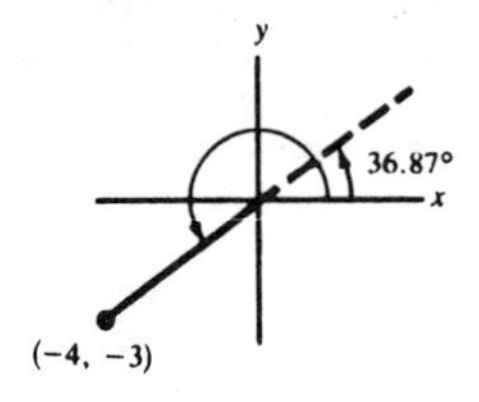

$\tan 36.87° = 0.75 = \tan 216.87°$

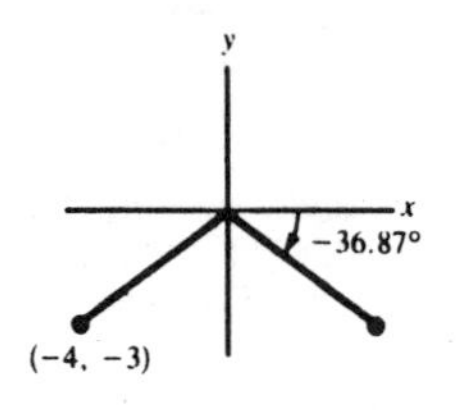

$\sin (-36.87°) = -0.600 = \sin 216.87°$

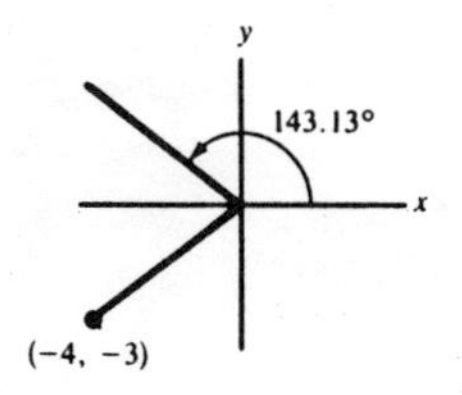

$\cos 143.13° = -0.800 = \cos 216.87°$

The double valued property of trigonometric functions may also be seen by examination of the functions plotted against angle. For illustration, the sine function is illustrated below. Notice that two angles satisfy value $\sin\theta = 0.5$; 30° and 150°. If you want to know which angle has a sine of 0.5, you have two possibilities. Your calculator will only tell you one of these two. If the angle is not in the first quadrant, you may have to interpret the value displayed by the calculator.

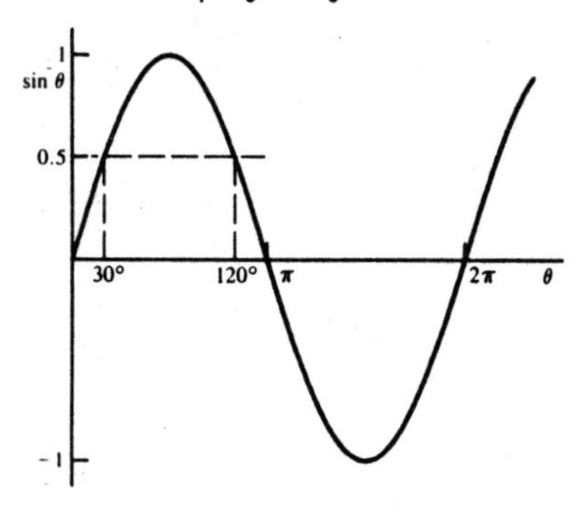

Practice Problems

50. The angle whose sine is 0.6538 is ______________.

51. arccos 0.7500 = ______________________________

52. $\tan^{-1} 7.365$ = ______________________________

Find the angles indicated.

53.

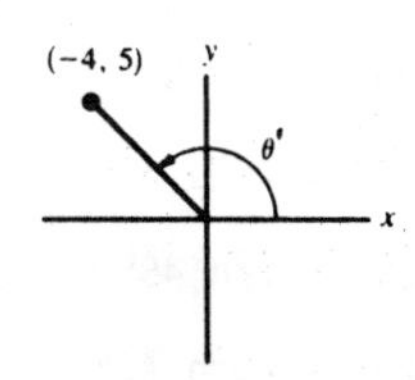

54.

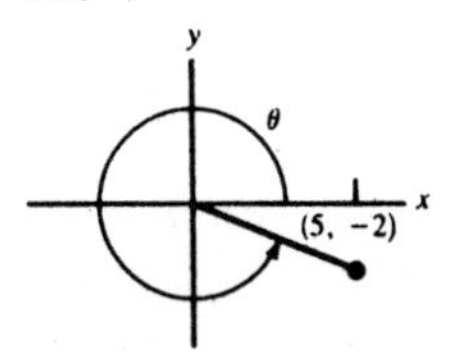

Answers to Practice Problems

1. 4
2. $\sqrt{7}$
3. $\sqrt{32} = 4\sqrt{2}$
4. $\sqrt{3}$
5. $\sqrt{a^2 - b^2}$
6. $\sqrt{x^2 - y^2}$

7. $\sqrt{x^2 + y^2}$

8. $\sqrt{p^2 - q^2}$

9. 60°

10. 144°

11. $180/\pi = 57.3°$

12. 22.5°

13. $\pi/180 = 0.0175$

14. $\pi/6$

15. $3\pi/2$

16. $\pi/12$

17. $\sin\theta = 1/\sqrt{3}$

$\cos\theta = 2/\sqrt{3}$

$\tan\theta = 1/2$

18. $\sin\theta = 4/5$

$\cos\theta = 3/5$

$\tan\theta = 4/3$

19. $\sin\theta = 1/\sqrt{5}$

$\cos\theta = 2/\sqrt{5}$

$\tan\theta = 1/2$

20. $\sin\theta = 1/\sqrt{3}$

$\cos\theta = \sqrt{2/3}$

$\tan\theta = 1/\sqrt{2}$

21. $\sin\theta = 3/5$

$\cos\theta = 4/5$

$\tan\theta = 3/4$

22. $\sin\theta = b/\sqrt{b^2 + c^2}$

$\cos\theta = c/\sqrt{b^2 + c^2}$

$\tan\theta = b/c$

23. $\sin\theta = q/\sqrt{q^2 + p^2}$

$\cos\theta = p/\sqrt{q^2 + p^2}$

$\tan\theta = q/p$

24. $\sin\theta = b/a$

$\cos\theta = \sqrt{a^2 - b^2}/a$

$\tan\theta = b/\sqrt{a^2 - b^2}$

25. $\cos 135° = -\cos 45°$

$\sin 135° = \sin 45°$

$\tan 135° = -\tan 45°$

26. $\sin 210° = -\sin 30°$

$\cos 210° = -\cos 30°$

$\tan 210° = \tan 30°$

27. $\tan 300° = -\tan 60°$

$\sin 300° = -\sin 60°$

$\cos 300° = \cos 60°$

28. $\cos 225° = -\cos 45°$

$\tan 225° = \tan 45°$

$\sin 225° = -\sin 45°$

29. $\sin\theta = -4/5$

$\cos\theta = -3/5$

$\tan\theta = 4/3$

30. $\sin\theta = -5/\sqrt{61}$

$\cos\theta = 6/\sqrt{61}$

$\tan\theta = -5/6$

31. $-\sin 30°$

32. $-\cos 30°$

33. $-\tan 60°$

34. $\cos(\pi/2)$

35. $\sin 30°$

36. $\cos(\pi/6)$

37. 60°

38. 45°

39. 20°

40. 75°

41. 2°

42. 33°

43. $\cos 30° = \sqrt{0.75}$; $\tan 30° = \sqrt{1/3}$

44. $\sin\beta = \sqrt{0.91}$; $\tan\beta = \sqrt{10.1}$

45. $\sin\lambda = \sqrt{1/1.25}$; $\cos\lambda = \sqrt{1/5}$

46. -75°

47. $\sin 405° = \sin 45°$

48. $\cos 90°$

49. $\sin 240°$

50. 40.8° or 139.2°

51. 41.4° or 318.6°

52. 82.3° or 262.3°

53. 129°

54. 338°

Exercises

Find the unknown side of the right triangles.

1.

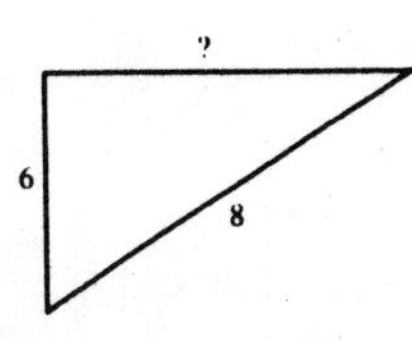

2.

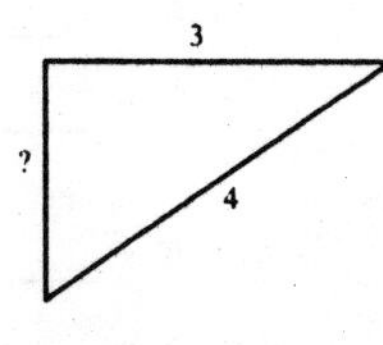

3.

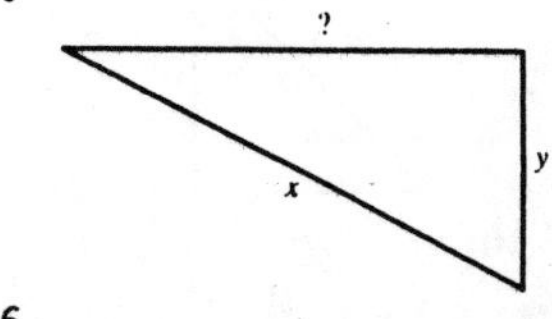

4.

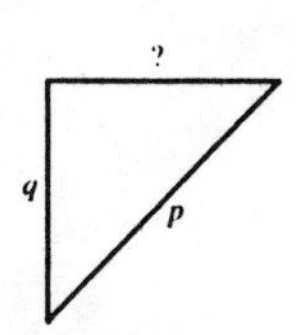

5.

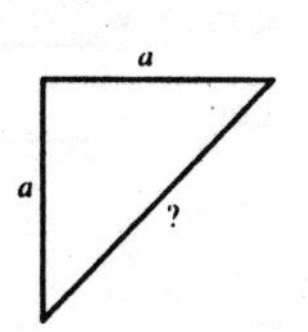

6.

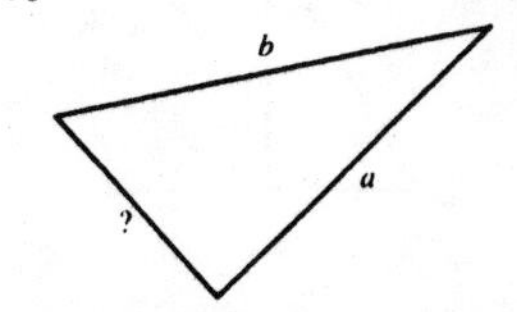

Convert from degrees to radians and <u>vice versa</u> as indicated.

7. $3\pi/4$ radians = __________ °

8. 330° = __________ radians

9. 315° = __________ radians

10. $5\pi/9$ radians = __________ °

11. 45° = __________ radians

12. $\pi/6$ radians = __________ °

13. $5\pi/6$ radians = ______ degrees

14. $5\pi/17$ radians = __________ °

Find the trigonometric functions for the following right triangles below.

15.

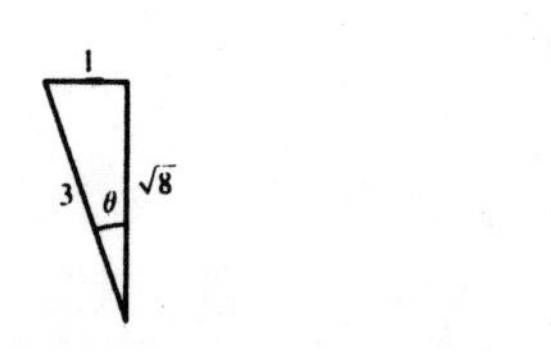

sin θ = ________________

cos θ = ________________

tan θ = ________________

16.

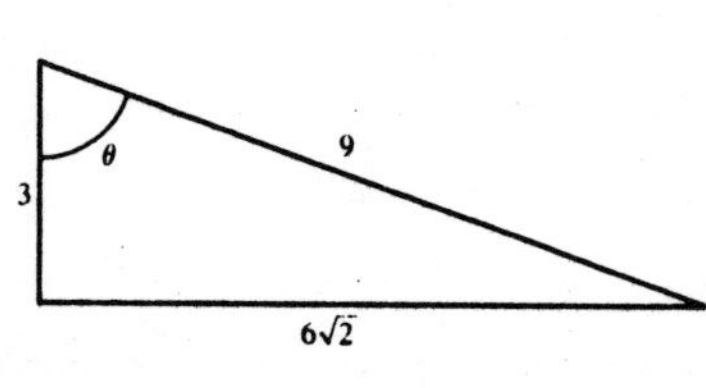

sin θ = ________________

cos θ = ________________

tan θ = ________________

17.

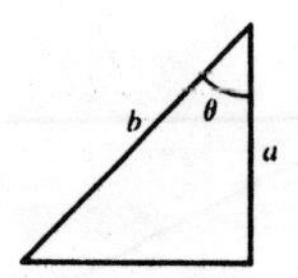

sin θ = ________________

cos θ = ________________

tan θ = ________________

18.

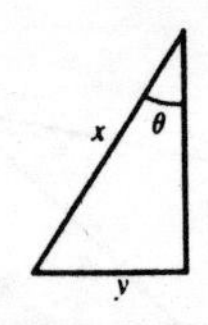

sin θ = ________________

cos θ = ________________

tan θ = ________________

19.

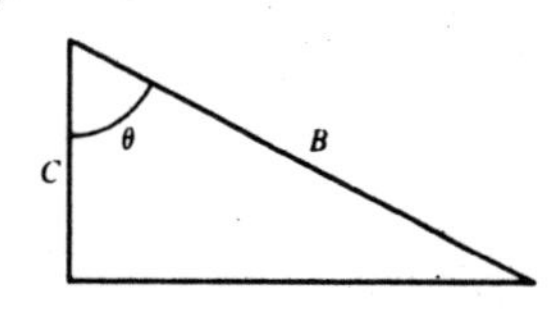

$\sin \theta =$ ____________________

$\cos \theta =$ ____________________

$\tan \theta =$ ____________________

20.

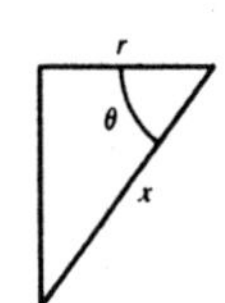

$\sin \theta =$ ____________________

$\cos \theta =$ ____________________

$\tan \theta =$ ____________________

Express the trigonometric function in terms of an angle in the first quadrant. If not identified by a symbol, the argument of the angle is in radians.

21. $\sin 210° =$ __sin ____________

22. $\cos 312° =$ __cos ____________

23. $\cos (5\pi/6) =$ __cos __________

24. $\sin (5\pi/4) =$ __sin __________

25. $\sin (7\pi/12) =$ __sin ________

26. $\cos (-\pi/6) =$ __cos __________

27. $\cos 210° =$ __cos ____________

28. $\cos (-70°) =$ __cos __________

29. $\sin -(3\pi/4) =$ __sin ________

30. $\cos (-5\pi/2) =$ __cos ________

31. On the sine function reproduced below, identify the angles 30°, 90°, 315°, $2\pi/3$ radians, $11\pi/6$ radians, $3\pi/2$ radians, 225°, and $13\pi/6$ radians.

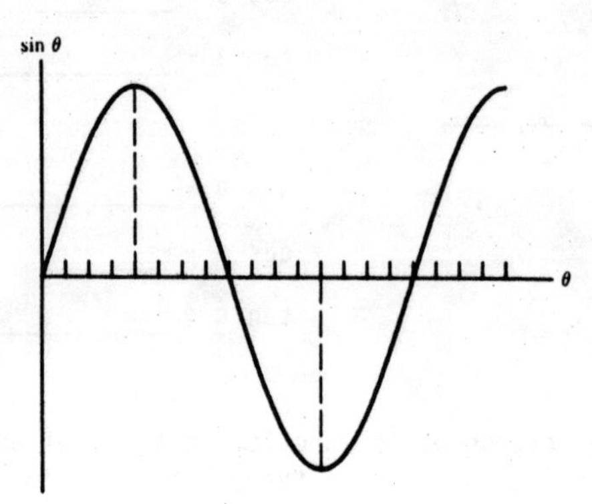

32. On the cosine function reproduced below, identify the angles $7\pi/4$ radians, $7\pi/6$ radians, 60°, 240°, $2\pi/3$ radians, 270°, $11\pi/6$ radians and 135°.

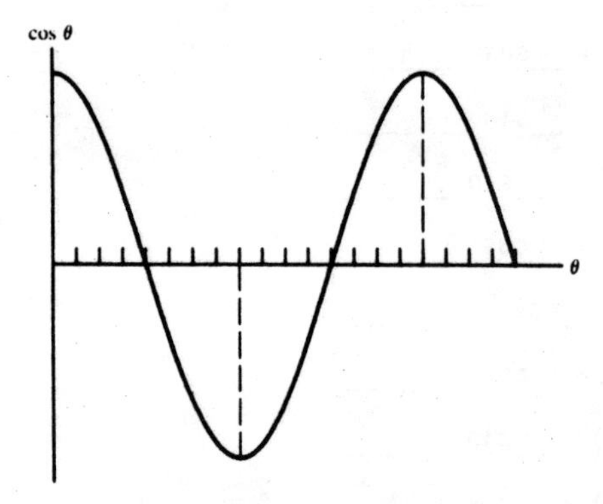

Express the trigonometric functions below in terms of an angle in the first quadrant.

33. sin (-40°)

34. cos (-120°)

35. tan 313°

36. sin 120°

37. tan 240°

38. cos $(7\pi/6)$

39. tan $(13\pi/4)$

40. sin $(16\pi/3)$

41. cos $(-8\pi/3)$

Express the following in terms of the complementary angle.

42. $\sin 60° = \cos$ __________

43. $\cos 15° = \sin$ __________

44. $\sin 1° = \cos$ __________

45. $\cos 5° = \sin$ __________

Use the relations $\cos \theta = \sin (\theta + 90°)$ and $\sin \theta = \cos (\theta - 90°)$ to convert the expressions below.

46. $\cos 60° = \sin$ __________

47. $\sin 195° = \cos$ __________

48. $\cos 312° = \sin$ __________

49. $\sin 215° = \cos$ __________

Find the angle indicated.

50. $\sin^{-1}(-0.333) =$ __________

51. $\cos^{-1}(-0.444) =$ __________

52. $\arctan (-3.225) =$ ______

53. $\arccos (-0.500) =$ ______

54. $\theta =$ __________°

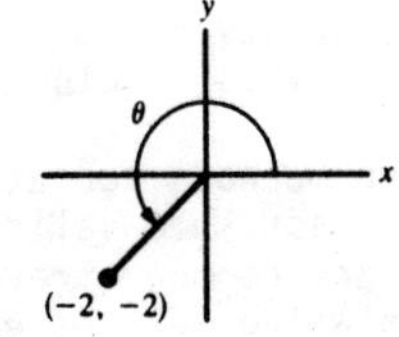

55. $\theta =$ __________°

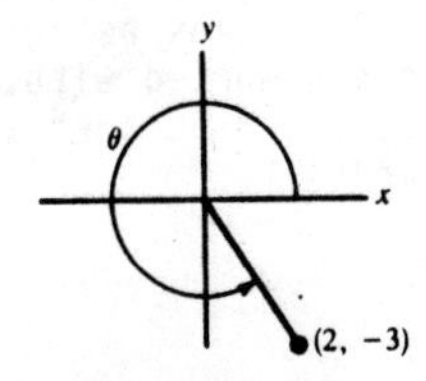

12 Word Problems

Word problems have been used in the preceding chapters wherever they could be worked into the material. However, many students of physics find word problems especially difficult, and it may be helpful to separate them out as a topic in their own right. If you feel discomfort when presented with a word problem, it will be worthwhile for you to work through this chapter.

The practice problems and exercises of this chapter may appear to have little to do with physics. At this stage, where you are reviewing your mathematics in preparation for a physics course, the important point is to develop a procedure for addressing problems. In any physics course, the problems are as much word problems as physics problems, in spite of the rumor you may have heard about physics that "all you need to know is which formula to use."

In this chapter you will be given a procedure broad enough to apply to a variety of different problem types, including those found in the context of a physics course.

A variety of step-by-step methods for attacking word problems have been suggested. All have at least some validity and none will work 100% of the time. Even so, there are common threads to the different procedures, and there is an accumulating body of evidence on how people learn problem solving techniques. The steps below appear to work for most of the students that the author has worked with. As you practice and improve your word problem skills, you may find that you develop your own techniques to fit your particular style.

12.1 GENERAL PROCEDURES

1. Read the problem. This advice may sound trite, but one dramatic difference between novice and expert problem solvers is that the novice grabs a pencil and starts to write while the expert is still reading the problem. There is no magic in the tip of your pencil and nothing is going to come out of it unless you know where you are heading. Put the pencil down until you understand what the problem is asking.

2. Read the problem again. If there is any doubt as to whether you understand the problem, pick up your pencil and write out the question asked, but in your own words (for example, "I am asked to find the position of the particle after ten seconds"). Do not use any symbols or equations. This step is to make sure you can express what you are looking for in a familiar language. If you can't express it in words, you can't hope to express it in symbols. You might also write out what you are given. You are not trying to solve the problem at this stage; you are making sure you understand it.

3. If appropriate, make a drawing. This does several things. First, it helps you visualize what is going on. Your mind is given a reference to understand the conditions stated. Second, it gives you some time to let your subconscious come to grips with the problem.

4. Examine your drawing. If it is sloppy or unclear, you probably don't understand the problem. Do another drawing on a clean piece of paper.

5. Write out the conditions of the problem in words. The expressions you are now writing will be a supplement to what you wrote in step 2 above. Leave space below each expression to come back later and put symbols. This step is to be part of a process of translating the words of the problem into symbolic representation.

6. Make a glossary or dictionary, using symbols and units. For example, you might have to find an area. Your glossary might include the entry:

A = area in meters squared.

The items you put in your glossary should certainly include all the information given, but it may eventually have to be expanded to include quantities calculated from the givens. Leave yourself some space.

7. Using the symbols of your glossary, translate each verbal expression from steps 3 and 6 into equations.

8. If it is obvious how the equations will be used to solve the problem, you are home free. Solve the equations. If it is not obvious, then work backward from the question. For example, if you are to find area, you must know length and width, or base and height, or radius, or some geometric factor. From which of the givens can this information be obtained. This may not be obvious unless you write out what you are given. Remember, every problem must contain adequate information or there is no solution.

As an example of how to apply these steps, consider the following problem:

> A rectangular plot of ground is four times as long as it is wide. If 1800 feet of fence is required to enclose the plot, what is the area of the plot?

1. Read the problem again. Are you sure you know what you are being asked to find? Write out what you know about the problem in words. For example:

> "I am to find the area of the rectangular plot. Area of a rectangle is length times width. I do not know either the length or the width, but I am told that the length is four times the width. I must find length and width, using the given information that the total perimeter is 1800 feet."

2. At this point, you may wish to draw a rectangle, roughly to scale.

3. Write out givens and see if they can be related to what you need to answer the problem.

> Area equal length times width (answer).
>
> Length is four times width (given).
>
> Circumference is 1800 feet (given).

Examine the drawing.

> Perimeter is width plus length plus width plus length, or twice the width plus twice the length.

Notice how you have multiple relations about length and width.

> Length is four times width.
>
> Perimeter is 1800 feet is twice width plus twice length.

4. Now you are ready to translate your expressions into equations. First, make a glossary.

> length = L (feet)
>
> width = w (feet)
>
> area = A ($feet^2$)

Try translating.

Length is four times width.

$L = 4w$.

Circumference is twice width plus twice length.

1800 feet = $2w + 2L$.

Now you have two equations in two unknowns. When you solve these, L and w will have units of feet.

$L = 4w$,

1800 feet = $2w + 2L$.

How you solve these is up to you (return to Chapter 8). One way is to substitute the first into the second.

1800 feet = $2w + 2\ (4w) = 2w + 8w = 10w$,

$w = 180$ feet.

So the width is 180 feet. The length is four times this, so $L = 4w$ becomes

$L = 720$ feet.

Finally, area = Lw = (180 feet)(720 feet) = A = 129,600 feet^2.

Try working the following problem on your own before referring to the solution.

Problem: Sue is five times as old as Joe. In five years, Sue will be three times as old as Joe. What are their ages now?

WORK THE PROBLEM IN THE SPACE BELOW:

Solution: Let S equal Sue's age now, and J equal Joe's age now. The statement, "Sue is five times as old as Joe" translates as

$$S = 5J.$$

Since Sue will be S + 5 and Joe will be J + 5 in five years, the statement, "In five years, Sue will be three times as old as Joe," becomes
Sue's age in 5 years = 3 times Joe's age in five years.

$$S + 5 \qquad = \qquad 3(J + 5)$$

The equations to solve are

$S = 5J$	$S = 5J$	$S = 5J$
$(S + 5) = 3(J + 5)$	$S + 5 = 3J + 15$	$S = 3J + 10$

Substituting S = 5J from the first equation into the second equation gives:

$$5J = 3J + 10$$

$$2J = 10$$

$$J = 5 \text{ (Joe's age now)}$$

$$S = 5J = 25 \text{ (Sue's age now)}$$

In five years, Sue will be 30 and Joe will be 10, or Sue will be three times as old as Joe.

A third example is given below. Again, do your best to work the problem before using the solution.

Problem: Dick and Harry work in the same plant assembling flim-flams. Working alone, Dick can assemble 12 flim-flams per hour, while Harry working alone can assemble 8 flim-flams per hour. How long will it take the two of them working together to assemble 100 flim-flams?

WORK THE PROBLEM IN THE SPACE BELOW:

Solution: Start by putting the question in words and work toward an equation in symbols.

1. The number of units assembled will be equal to the number assembled by Dick plus the number assembled by Harry.

2. Total number assembled = number assembled by Dick + number assembled by Harry.

You are given the rate (per hour) that Dick and Harry assemble units, so express the number of units assembled by each in terms of rate times time.

3. Total = Dick's hourly rate times hours worked + Harry's hourly rate times hours worked.

You are looking for the time Dick and Harry will have to work together, which means each will work the same time. Let some variable be equal to this unknown time.

T = number of hours Dick and Harry will have to work to assemble 100 units.

Equation (3) becomes

4. Total = Dick's hourly rate times T + Harry's hourly rate times T.

In the problem you are given the following:

Total units = 100.

Dick's hourly rate = 12 per hour.

Harry's hourly rate = 8 per hour.

In T hours, Dick will assemble $12T$ units and Harry will assemble $8T$ units. Together they will assemble $20T$ units. Using this information, Equation (4) becomes

5. Total = (12 units/hour)T + (8 units/hour)T.

Solving for T:

$$100 = 20T/\text{hour},$$

$$T = 5 \text{ hours}.$$

Check: In 5 hours Dick will assemble (5)(12) = 60 units and Harry will assemble (5)(8) = 40 units. 60 + 40 = 100 units.

Aside from a mathematical check, examine the answer to see if it seems reasonable. Harry would require 100/8 = 12.5 hours by himself and Dick would require 100/12 = 8.3 hours. The time required by both would be less than the time of either working alone, because each unit assembled by one worker reduces the number of units left to be assembled. Dick is the faster worker. If Harry worked as fast, the time would be exactly half that of Dick's alone or 4.2 hours. Since Harry isn't as fast, the time will be just a little longer than 4.2 hours. The 5 hour requirement is reasonable.

Practice Problems

1. A grandfather is five times as old as his granddaughter. Their combined ages total 72 years. How old is the grandfather?

2. A man is twice as old as his son. The combined age of the two is 54 years. How old is the son?

3. The sum of two numbers is 40 and their difference is 10. What are the two numbers?

4. Beth has $250 in her account. If she makes an initial withdrawal of $50 and then takes out $5 each day, starting with the second day, how long will it take to empty the account?

5 A rectangular plot of ground has a length which is twice its width. If 1200 feet of fence is required to enclose the plot completely, what is the width?

6. The difference of two numbers is 12 and their product is 64. What are the numbers?

7. Sue and Madge are painters. In a motel complex, all of the rooms are the same size. Working alone, Sue can paint three rooms a day. Working alone, Madge can paint five rooms a day. How long will it take Sue and Madge working together to paint 40 rooms?

8. A rectangular wall has a length three times its height. If the total area of the wall is 432 square feet, what is the perimeter?

9. A contractor uses a pump to remove the water from a ditch. He has a pump that will empty the ditch in four hours. He rents a second pump and the two pumps empty the ditch in three hours. How long would it take for the second pump alone to empty the ditch?

10. The sum of two numbers is 17 and their product is 72. What are the numbers?

Answers to Practice Problems

1. 60 years
2. 18 years
3. 25 and 15
4. 41 days
5. 200 feet
6. 4 and 16
7. 5 days
8. 96 feet
9. 12 hours
10. 9 and 8

Exercises

1. A 1200 ft^2 rectangular plot is three times as wide as it is long. What are the dimensions of the plot?

2. Dorothy is three years older than Lisa. The sum of their ages is 21 years. How old is Lisa?

3. Three years ago Tom was twice as old as Joe. Today Tom is seven years older than Joe. How old is Tom today?

4. The sum of two numbers is 13 and their product is 42. What are the numbers?

5. The difference of two numbers is 10 and their product is 96. What are the two numbers?

6. A rectangular plot has a length 2 1/2 times the width. The area of the plot is 4000 ft^2. What is the perimeter?

7. Joe can paint a certain area in six hours while Edna can paint the same area in four hours. How long does it take the two of them working together to paint the area?

8. A farmer can cut and bale the hay in a large field in six hours. The farmer's son, using a second machine, can cut and bale the hay in eight hours. How long will it take the two of them working together to cut and bale the hay?

9. Oscar and Sue usually work together as painters. Together they can paint a large room in two hours. Oscar was sick, and Sue, working alone, required five hours to paint the room. How long would it have taken Oscar to paint the room by himself?

10. Five years ago Lynne was three times as old as Horace. In five more years she will be twice as old as he will be. What are their ages today?

Practice Diagnostic Test

The test that follows covers the topics you will be using in the physics course. The test is diagnostic in that the questions you miss indicate the problem areas in your mathematics knowledge. Set aside a one hour period to take the test. Work the problems, marking your answer choice. After completing the test, consult the answers and diagnostic at the end.

As you take the test, be aware of the time factors that will be very real to a physics course. If you require more than two minutes to answer any question, then you should review and practice with that topic regardless of whether you could eventually arrive at the answer. In an examination your time will severely limited. Practice with and polish up your mathematics skills before the physics examination.

DIAGNOSTIC TEST FOR GENERAL PHYSICS

1. Given $s = 4$, $t = -5$, and $x = 9$. What is $(x - 2s - t)$?

    a. -4
    b. -6
    c. +4
    d. +6
    e. +12

2. Solve for t in the expression $xt + y = x$.

    a. $1 - y$
    b. $(x - y)/x$
    c. $x - xy$
    d. $-y + 1$
    e. $1 - y$

3. $\sqrt{100 - 64} = ?$

    a. 2
    b. 4
    c. 6
    d. 8
    e. 10

4. Given $yt^2 + zt + x = 0$. If x, y, and z are constant, then $t = ?$

    a. $\dfrac{-y \pm \sqrt{z^2 - 4x}}{2z}$
    b. $\dfrac{-x \pm \sqrt{y^2 - 4xy}}{2y}$
    c. $\dfrac{-z \pm \sqrt{y^2 - 4xz}}{2x}$
    d. $\dfrac{-x \pm \sqrt{y^2 - 4z}}{2z}$
    e. $\dfrac{-z \pm \sqrt{z^2 - 4xy}}{2y}$

5. Solve for y as a function of x from the equations below.

    $x = 3t$

    $y = 2t - 5$

    a. $y = (2x/3) - 5$
    b. $y = (2x - 3)/5$
    c. $y = (3x/2) - (5/3)$
    d. $y = 2x - (5/3)$
    e. $y = 2x - (3/5)$

6. Solve for x as a function of y from the equations below.

$x = 12z^2$

$y = 3z$

a. $x = 4y^2$

b. $x = 36y^2$

c. $x = 4y^2/3$

d. $x = 108y^2$

e. $x = y^2/4$

7. What is the slope of the line at x = 4?

a. +0.75
b. +1.5
c. -0.75
d. -1.5
e. 6

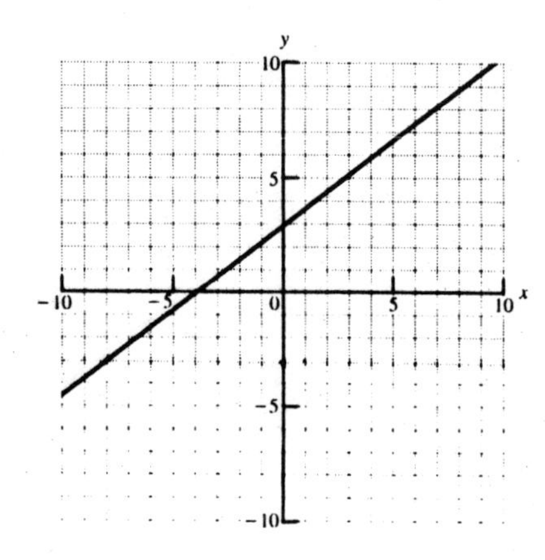

8. 3/20 + 5/12 = ?

a. 1/4
b. 2/15
c. 1/2
d. 17/30
e. 9/16

Questions 9, 10, and 11 refer to the triangle illustrated.

9. sin α = ?

a. 3/5
b. 3/4
c. 4/3
d. 4/5
e. 5/3

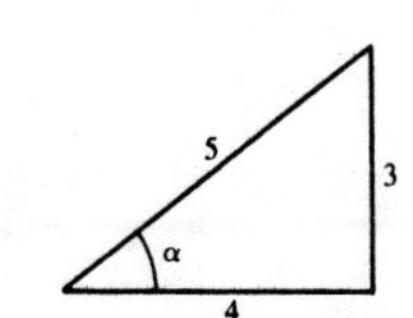

10. $\cos \alpha = ?$

a. 3/5
b. 3/4
c. 4/3
d. 4/5
e. 5/4

11. $\tan \alpha = ?$

a. 3/5
b. 3/4
c. 4/3
d. 4/5
e. 5/3

12. What is the value of the unknown side in the right triangle illustrated?

a. $\sqrt{7} - \sqrt{5}$
b. $\sqrt{24}$
c. 4
d. 2
e. $\sqrt{2}$

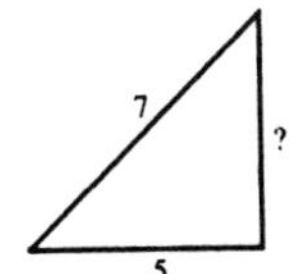

13. Which of the following is correct for an angle of 300°?

a. $\cos 300° = -\cos 30°$
b. $\cos 300° = \cos 30°$
c. $\cos 300° = -\cos 60°$
d. $\cos 300° = \cos 60°$
e. $\cos 300° = 1/\cos 30°$

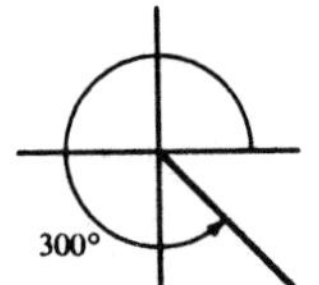

14. How many radians are there in 150°?

a. $3\pi/2$
b. $3\pi/4$
c. $3\pi/5$
d. $4\pi/3$
e. $5\pi/6$

15. $(3\times10^{-6})\times(-2\times10^{4}) = ?$

a. 1×10^{-2}
b. -6×10^{10}
c. 6×10^{-10}
d. -6×10^{-2}
e. -6×10^{-10}

16. David is 14 years older than Fred. Three years ago, David was three times as old as Fred was then. How old is Fred now?

a. 8 years
b. 10 years
c. 11 years
d. 17 years
e. 237 years

17. What is $(ax^2 - dx - e)$ if $a = -2$, $d = +3$, $e = -5$, and $x = -4$?

a. -15
b. -39
c. +25
d. -47
e. +47

18. What are the roots of the equation $x^2 - 7x + 12 = 0$?

a. $x = 12; x = -7$

b. $x = 19; x = 5$

c. $x = 3; x = 4$

d. $x = \sqrt{19}; x = \sqrt{5}$

e. $x = -19; x = -5$

19. Solve for y as a function of x from the equations below when a, b, and c are constant.

$x - a = bt$

$y - b = abt - c$

a. $y = ax - a^2 - c + b$

b. $y = abx = c + b$

c. $y = abx - a^2b - c$

d. $y = abx - c$

e. $y = ax - a - (c + b)$

20. Given $ax^2 + b = c$, and $aby = ac$, where a, b, and c are constant, find y as a function of x.

a. $y = (ax^2 + b)/ab$

b. $y = (ax^2 + b)/b$

c. $y = (ax^2 + b)/abc$

d. $y = (ax^2 + b - c)/ab$

e. $y = (ax^2 + b + c)/ab$

21. What is the slope of the straight line through the points $(x = 3, y = 5)$ and $(x = -5, y = 9)$?

a. $-(1/2)$
b. $+(1/7)$
c. $1/7$
d. $4/7$
e. $-(4/7)$

22. $\dfrac{12 \times 10^{-15}}{3 \times 10^{-5}} = ?$

a. 4×10^{3}

b. 9×10^{-10}

c. 4×10^{-20}

d. 4×10^{-10}

e. 9×10^{-20}

23. For the triangle at the right, what is tan θ?

a. $(\sqrt{p^2 + q^2})/p$

b. $(\sqrt{p^2} - \sqrt{q^2})/q$

c. $(\sqrt{p^2 - q^2})/q$

d. $q/\sqrt{p^2 + q^2}$

e. $p/\sqrt{p^2 - q^2}$

24. What is sin θ for the triangle illustrated?

a. $(\sqrt{a^2 - b^2})/a$

b. $\sqrt{a^2 + b^2}/a$

c. $b/\sqrt{a^2 + b^2}$

d. $b/\sqrt{a^2 - b^2}$

e. $a/\sqrt{a^2 - b^2}$

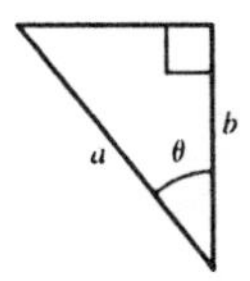

25. Which of the following is true for a 250° angle?

a. $\tan 250° = -\tan 20°$
b. $\tan 250° = \tan 20°$
c. $\tan 250° = -\tan 70°$
d. $\tan 250° = \tan 70°$

e. $\tan 250° = \dfrac{1}{\tan 20°}$.

26. What is the unknown side in the triangle illustrated?

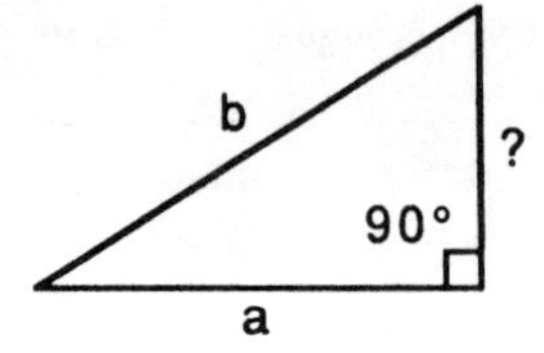

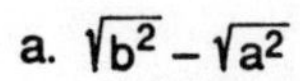

a. $\sqrt{b^2} - \sqrt{a^2}$

b. $(\sqrt{b - a})^2$

c. $\sqrt{b^2 - a^2}$

d. $\sqrt{b} - \sqrt{a}$

e. $\sqrt{a^2 + b^2}$

27. $\frac{5\pi}{18}$ radians = _____ degrees?

a. 30°
b. 18°
c. 50°
d. 90°
e. 5.5°

28. In the right triangle illustrated, what is r in terms of p, q and cos θ?

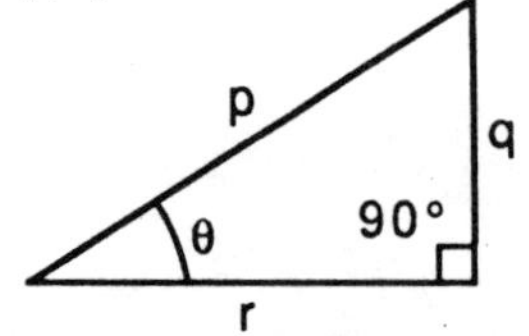

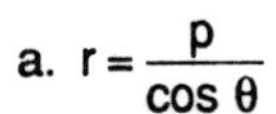

a. $r = \dfrac{p}{\cos\theta}$

b. $r = p\cos\theta$

c. $r = \dfrac{q}{\cos\theta}$

d. $r = q\cos\theta$

e. $r = \sqrt{p^2 - q^2}\cos\theta$

29. What is the value of the unknown side in the triangle illustrated?

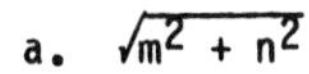

a. $\sqrt{m^2 + n^2}$

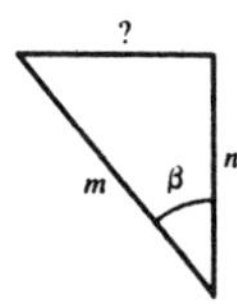

b. $\sqrt{m} - \sqrt{n}$

c. $\sqrt{m} + \sqrt{n}$

d. $\sqrt{m^2 - n^2}$

e. $m - n$

30. What values of x satisfy the equation $4x^2 - 2x - 3 = 0$?

a. $x = (1 \pm \sqrt{13})/4$

b. $x = (1 \pm \sqrt{3})/2$

c. $x = 2 \pm 3\sqrt{12}$

d. $x = (1 \pm 3\sqrt{3})$

e. $x = (1 \pm 3\sqrt{3})/2$

31. Joe and Ed are welders for a pipeline company. A certain job would take Joe, working alone, 16 hours to complete. Ed could do the same job in 12 hours. How long will it require for the two of them working together to complete the job?

a. 14 hrs

b. 9.3 hrs

c. 8 hrs

d. 6.9 hrs

e. 4.6 hrs

32. $1.63 \times 10^5 + 3.14 \times 10^7 = ?$

a. 3.16×10^7

b. 4.77×10^5

c. 4.77×10^7

d. 1.66×10^5

e. 1.66×10^7

33. $\frac{xy}{z^2} + \frac{y}{x^2z} = ?$

a. $\frac{y(x^2 + z^2)}{x^2z}$

b. $\frac{y(x^3 + z)}{x^2z^2}$

c. $\frac{xy(z^2 + x)}{x^2z^3}$

d. $\frac{y(x + 1)}{x^2z^2}$

e. $\frac{y(x + 1)}{z(z + x^2)}$

34. $\frac{x^3y^2}{rz} + \frac{xr}{y^3z} = ?$

a. $\frac{x^4}{yz^2}$

b. $\frac{x^2y}{z^2r^2}$

c. $\frac{x^2y^5}{r^2}$

d. $\frac{x^2r^2}{yz^2}$

e. $\frac{x^2r^2}{yz^2}$

35. $\frac{a - 1}{a} + a = ?$

a. a

b. $a - 1$

c. $\frac{a^2 + a - 1}{a}$

d. $\frac{a^2 - a + 1}{a}$

e. $\frac{2a - 1}{a}$

36. What is the (approximate) slope of the curve at $x = 3$?

a. -1

b. +1

c. -0.5

d. +0.5

e. +2

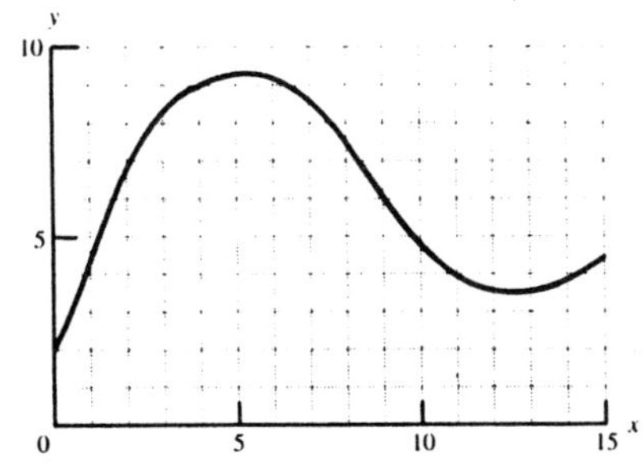

37. What are the values of x that satisfy the equation

$ax^2 + (d - ab)x - db = 0$?

a. $x = -d/a$; $x = b$

b. $x = d$; $x = b/a$

c. $x = a + d$; $x = -d/b$

d. $x = db/a$; $x = 1/(a + d)$

e. $x = (a + d)/a$; $x = db/a$

38. $1.35 \times 10^{-4} + 3.63 \times 10^{-6} = ?$

a. 4.98×10^{-4}

b. 4.98×10^{-6}

c. 1.39×10^{-4}

d. 3.64×10^{-6}

e. 3.64×10^{-4}

39. $\sqrt{64} + \sqrt{25} = ?$

a. $\sqrt{89}$

b. 13

c. $\sqrt{13}$

d. $\sqrt{39}$

e. 89

40. $\frac{a}{a + b} - \frac{1}{b} = ?$

a. 0

b. 1/b

c. 2/b

d. (ab - a - b)/b(a + b)

e. (a - b)/b(a + b)

41. $\frac{12}{25} \div \frac{7}{15} = ?$

a. 84/15
b. 36/35
c. 15/84
d. 35/36
e. 60/21

42. What is the slope of the line which goes through the origin and the point x = (-2, y = 4)?

a. 2
b. 1/2
c. -1/2
d. -2
e. -8

43. Solve for x in the expression 2(x - 4) + 3 = 6x.

a. 4
b. -5/6
c. 11/8
d. 5/6
e. -5/4

Answers to Diagnostic Test

The answers to the diagnostic test are given below. However, this is not a test where a "good" score means that you are ready for the physics course. All of the skills and knowledge on this test will eventually be required of you. Check off the questions you missed and then refer to the diagnostic that follows.

1. d. $+6$
2. b. $(x - y)/x$
3. c. 6
4. e. $(-z \pm \sqrt{z^2 - 4xy})/2y$
5. a. $(2x/3) - 5$
6. c. $4y^2/3$
7. a. $+0.75$
8. d. $17/30$
9. a. $3/5$
10. d. $4/5$
11. b. $3/4$
12. b. $\sqrt{24}$
13. d. $\cos 60°$
14. e. $5\pi/6$
15. d. -6×10^{-2}
16. b. 10 years
17. a. -15
18. c. $4, 3$
19. a. $y = ax - a^2 - c + b$
20. b. $(ax^2 + b)/b$
21. a. $-1/2$
22. d. 4×10^{-10}
23. c. $\sqrt{p^2 - q^2}/q$
24. a. $\sqrt{a^2 - b^2}/a$
25. d. $\tan 70°$
26. c. $\sqrt{b^2 - a^2}$
27. c. $50°$
28. b. $p \cos \theta$
29. d. $\sqrt{m^2 - n^2}$
30. a. $(1 \pm \sqrt{13})/4$
31. d. 6.9 hrs
32. a. 3.16×10^7
33. b. $(x^3y + yz)/x^2z^2$
34. c. x^2y^5/r^2
35. c. $(a - 1 + a^2)/a$
36. b. $+1$
37. a. $-d/a$; b
38. c. 1.39×10^{-4}
39. b. 13
40. d. $(ab - a - b)/b(a + b)$
41. b. $36/35$
42. d. -2
43. e. $-5/4$

Diagnostic For Math Test

Each of the topics below is referenced to questions from the diagnostic test. In the questions columns below, circle the number of each question you missed or omitted.

If you missed both questions for any topic, this indicates that you need to learn that topic. On the other hand, you most probably understand and are readily able to work with those topics for which you correctly answered both questions. For any topic which you correctly answered only one question, you may only need to review.

The last three listings have to do with more sophisticated skills than simply knowing a definition or a mechanical operation. In this analysis, there is a first requirement that a particular skill be known. If you miss the questions related to these topics, you need to practice.

	Topic	Questions	
1.	Clearing equations	1	17
2.	Linear algebra	2	43
3.	Square roots	3	39
4.	Quadratic equation	4	30
5.	Linear parametric equations	5	19
6.	Parametric equations - one a quadratic	6	20
7.	Graphs	7	36
8.	Adding fractions	8	33
9.	Sine function	9	24
10.	Cosine function	10	28
11.	Tangent function	11	23
12.	Pythagorean theorem	12	26
13.	Angles larger than 90°	13	25
14.	Degrees and radians	14	27
15.	Powers of 10	15	22
16.	Word problems	16	31
17.	Factoring quadratic equations	18	37
18.	Equation of straight line	21	42

19.	Adding powers of 10	32	38
20.	Dividing fractions	34	41
21.	Cancelling	35	40
22.	Two-step problems	10 and 12 correct, 28 incorrect 9 and 12 correct, 24 incorrect	
23.	Working with symbols	6 correct and 20 incorrect 10 correct and 28 incorrect 12 correct and 26 incorrect	
24.	Rotated figures	26 correct and 29 incorrect 9 correct and 24 incorrect	

Answers to Exercises

Chapter 1

1. 0.3
2. 0.3
3. 2.76
4. -0.7
5. 1.3
6. 20
7. 270
8. 20
9. 600,000
10. 94
11. 17.2

Chapter 2

1. 10^5
2. 10^3
3. 10^1
4. 10^{-3}
5. 10^{-1}
6. 10^1
7. 10^{-3}
8. 10^{-4}
9. 10^{-6}
10. 10^{-6}
11. 0.002
12. 230,000
13. 120,000,000
14. 12600

15. 5230
16. 71300
17. 162,300,000,000,000,000
18. 0.000124
19. 0.127
20. 0.0001536
21. 0.00012
22. 0.000003
23. 0.00000000004
24. 0.00000123
25. 100,000,000
26. 7,200,000,000,000,000,000
27. 0.00000024
28. 0.0000024
29. 0.00000000000006
30. 4000
31. 0.004
32. 800,000
33. 600,000
34. 0.0000000033 or 0.0000000031
35. 200000

Chapter 3

1. 5
2. 4
3. -1
4. 15
5. 17
6. 13
7. 5
8. -15
9. 10
10. -29
11. 23
12. -41
13. 5 drinks; 35 cents
14. 62.5 minutes (assuming race is exactly 10 miles)
15. 8.8 gallons

Chapter 4

1. 3/11
2. 5/9
3. 8/13
4. 18/25
5. $2b^2/7ac$
6. $2p^2\Omega/q^3v^2x$
7. (a - b)/2
8. a^2/z
9. 1
10. $5c^2/26ab^3$

11. $7a^5b^4/2d^2$

12. $5x^3/3qyz$

13. $[c(b^2 - c)]/[d(c - x^2)]$

14. $(x - y)^2/xz^2$

15. 81/125

16. 49/20

17. b^4/y^2z

18. $2a/3(b - 2)$

19. 71/175

20. 39/58

21. 7/48

22. -1/90

23. $2b/(a^2 - b^2)$

24. $[4x^2 + (x + 4)^2]/[2x(x + 4)]$

Chapter 5

1. $w = 7/2$
2. $b = -5$
3. $x = 12$
4. $x = 1/2$
5. $z = -14/3$
6. $p = 4$
7. $q = -1$
8. $(\alpha - \beta)/2$
9. $(\theta - s - 3)/\theta$
10. $(2 - y)/4$
11. $3z + 2y$
12. $(c - A)/(B - D)$
13. $(a + 4b + 5)/(2 + b)$
14. $(8t - \alpha)/t(6 - v_o)$
15. $(2a + 2b + ab)/(9a + 5b)$
16. $(24 + 2c + 4t)/(t + 15 + 5c)$
17. $(4a - b^2 - ab)/(b - 2b^2 + 4 + a)$
18. $Ay/(A^2 - 3y^2)$
19. -4
20. -63
21. -8/19
22. 24/5
23. $x - at - bt^2$
24. $(2s - 2r + at^2)/2t$
25. $2(v - u + vt)/t^2$
26. $(A + B + at^2)/t$
27. $(y + xyt^2 - x)/t$
28. 34
29. 8
30. $8
31. 22
32. 3
33. 12
34. -5
35. 4
36. 1.8 hours
37. 109
38. -8.9°

Chapter Six

1. $4\sqrt{2}$
2. $3\sqrt{3}$
3. $3\sqrt{2}$
4. $5\sqrt{2}$
5. $5\sqrt{10}$
6. $b\sqrt{a}$
7. $x\sqrt{xy}$
8. $A^2\sqrt{x}$
9. $2\omega^2\theta\sqrt{\omega\alpha}$
10. $pq^2\sqrt{2gp}$
11. 5
12. $\sqrt{13}$
13. $3\sqrt{3}$
14. $a^2\sqrt{b}$
15. $xy\sqrt{x}$
16. $p^2 + p$
17. $A^2\sqrt{B} + AB\sqrt{A}$
18. $6a^2x\sqrt{2x}$
19. $2\sqrt{x}$
20. $a/3(b - 1)$
21. $2\sqrt{3} - 3$
22. $8 + 3\sqrt{6}$
23. $(1 - y\sqrt{2})/y^2$
24. $(2p\sqrt{q} + q^2\sqrt{6q})/6p^2q$
25. $\sqrt{55}$
26. $\sqrt{a^2 - b^2}$
27. $\sqrt{6} + 3\sqrt{2} + 2 + 2\sqrt{3}$

Chapter Seven

1. $(1 \pm \sqrt{7})/2$
2. $(7 \pm \sqrt{41})/2$
3. 2
4. $-2 \pm \sqrt{6}$
5. $(-3 \pm \sqrt{21})/6$
6. $(1 \pm \sqrt{25})/2$
7. $(4 \pm \sqrt{10})/3$
8. $(-5 \pm \sqrt{37})/2$
9. -2; 1/2
10. $(-21 \pm \sqrt{481})/4$
11. -1; 2
12. 5; -3
13. 2; 3
14. -1; -3
15. $-4 \pm 2\sqrt{2}$
16. -1; -3
17. 1/2; -1
18. -1; -1/3
19. 4; -1
20. 1/3; -3/2
21. ±1/2
22. 1/4; -1

23. $[-B \pm \sqrt{B^2 - 4AC(D - C)}]/2A$

24. $[-q \pm \sqrt{q^2 + 4qp}]/2q$

25. $[a^2 - b^2 \pm \sqrt{(b^2 - a^2)^2 - 4ab(b - a - ab)}]/2ab$

26. $[\alpha(D + 2) \pm \sqrt{\alpha^2(D + 2)^2 + 4wD}]/2D$

27. $[b \pm \sqrt{b^2 - 4(a - 1)(a^2 - a - ab)}]/2(a - 1)$

28. $[-b(2a - 1) \pm \sqrt{b^2(2a - 1)^2 - 4a^2c^2}]/2ac$

29. $[-v_o \pm \sqrt{v_o^2 - 2a(p_o - p)}]/a$

30. $(y \pm \sqrt{y^2 + 12y})/6$

31. $(v_o \pm \sqrt{v_o^2 - 2a(x_o - x)})/a$

32. $[C \pm \sqrt{C^2 + 4D^2(C - D)}]/2D$

33. $(18 \pm 10\sqrt{15})/7$

34. $1 \pm \sqrt{2}$

35. 1/2; -2

36. $(1 \pm \sqrt{5})/2$

37. 1

38. 1

39. -1

40. $[(C - B) \pm \sqrt{(B - C)^2 + 4CB}]/2A$

Chapter Eight

1. $y = 4/3$; $x = 8/3$
2. $x = 5/2$; $y = -3/4$
3. $x = 11/8$; $y = 9/8$
4. $p = 6/7$; $q = 23/7$
5. $a = 7/5$; $b = -11/20$
6. $A = 18/5$; $C = 32/5$
7. $x = 8/11$; $y = -24/11$
8. $x = 26/7$; $y = -4/7$
9. $p = 5/3$; $b = 2/3$
10. $x = 0$; $y = 0$
11. $y = -7/4$; $x = 1/2$
12. $y = 12$; $p = 20$
13. $A = -2$; $z = 0$
14. $x = -18/31$; $y = -27/31$

15. $x = 5; y = 1/2$

16. $(x = -2/3; y = -17/3)$
$(x = -4; y = 1)$

17. $x = (-3 \pm 4\sqrt{15})/21$
$y = (3 \pm 3\sqrt{15})/21$

18. $(v_1 = -2; v_2 = 1)(v_1 = 2/5; v_2 = -13/5)$

19. $(p = 2; q = 1)(p = 6/11; q = 17/11)$

20. $(p = -5; q = 1)(p = 23/5; q = -7/5)$

21. $x = -4 \pm 4\sqrt{2}; t = -4 \pm 2\sqrt{2}$

22. $a = (4 \pm \sqrt{5})/11; b = (2 \pm 6\sqrt{5})/11$

23. $A = (24 \pm 2\sqrt{39})/35; B = (4 \pm 12\sqrt{39})/35$

24. $p = (1 \pm \sqrt{39})/2; q = (-3 \pm \sqrt{39})/6$

25. 10 boys and 20 girls

26. 3 and 12

27. 20 miles

28. 20 cents for bean and 25 cents for tomato

29. Joe is 12 and Maybelle is 6

30. 12 quarters and 15 dimes

31. 5 and 16

32. 2 inches x 3 inches

33. 6 apples and 12 oranges

34. 17 and 13

Chapter Nine

1. $12y + 2 = 5x$

2. $12y - 19 = 6t$

3. $42d - 48b = 23$

4. $6xy = 6y + 1$

5. $A(2Z - 1) = w - 2$

6. $V(y + 2) = U(x + 4)$

7. $R(x - 2) = 3BR + B - 3Xz$

8. $B(A - D^2)ac = A^2BDa + CDc - ACD^2$

9. $abZ(2R - 3) = 8BR(b + 1)$

10. $y(3B - A^2Bx - 2A) = 12B$

11. $2Ax = v^2 - V_o^2$

12. $b^2(c - 2) + b(c - 2) - 12c = 0$

13. $2x = 2v_0T + vT$

14. $30\Omega - 18q = 25q\Omega^2$

15. $B^3z = A^2Cx^2 - 2AC^2x + C^3 + AB^2$

16. $ABx = y^2 - 14y + 49 + BC$

17. $D(1 - C)^2a^2 - BC(1 - C)ab - B^2Db = 0$

18. $Rr + S = Rs - T$

19. $(B - C)^2Bb - ACa(B - C) = a^2A^2$

20. $A^2Ba - A^2B^2 = AT^2b + B$

21. $x = \pm 4\sqrt{(y + 4)}/3$

22. $\dfrac{4 \pm \sqrt{4 + 9y}}{2}$

Chapter Ten

1. $x = -0.67y - 1.7$

2. $v = 0.86t - 6$

3.

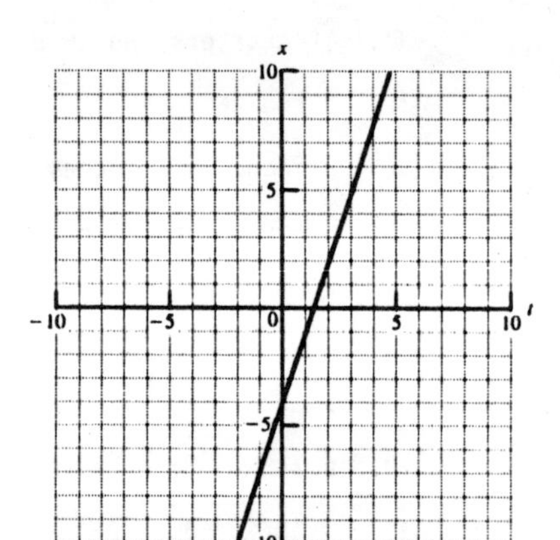

5.

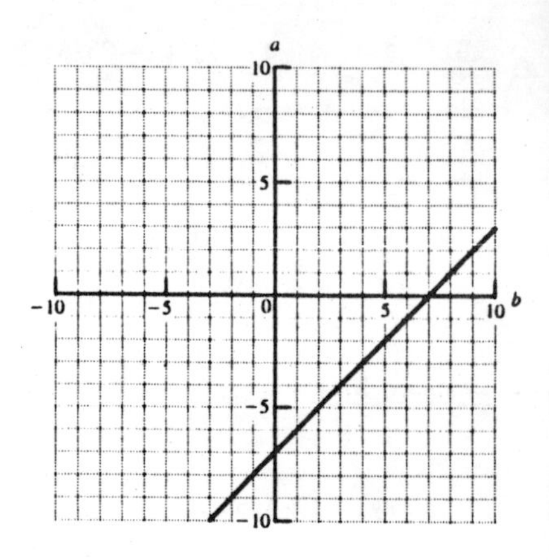

4.

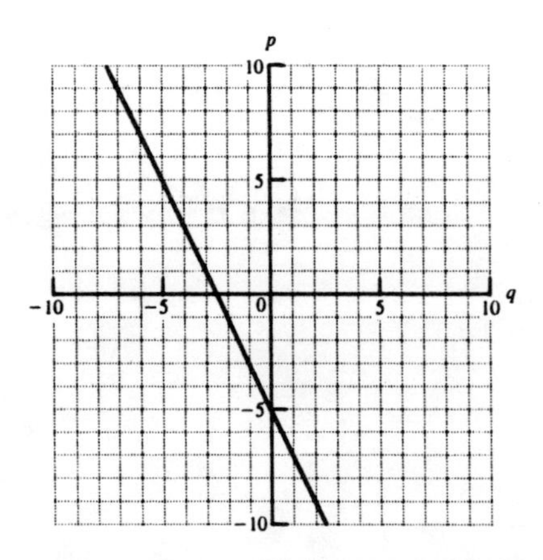

6.

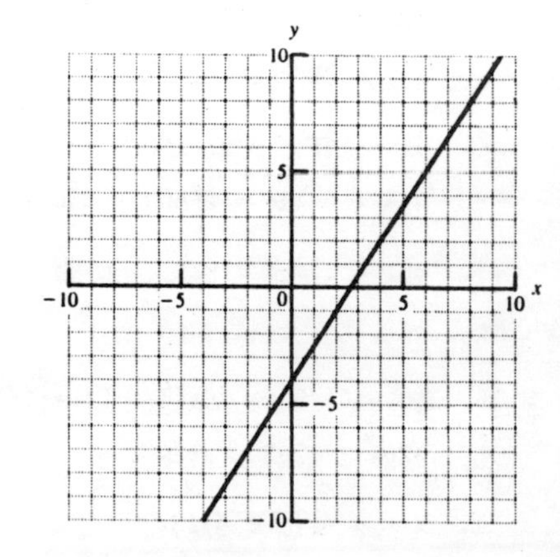

Chapter Ten (cont'd)

7. slope 2/3; intercept -42
8. slope -2; intercept -225
9. $t = 5.3, 15, 22.3, >40$
10. ≈ 2.2
11. ≈ 2
12. ≈ -2.2

Chapter Eleven

1. $\sqrt{28}$
2. $\sqrt{7}$
3. $\sqrt{x^2 - y^2}$
4. $\sqrt{p^2 - q^2}$
5. $a\sqrt{2}$
6. $\sqrt{a^2 - b^2}$
7. 135°
8. $11\pi/6$
9. $7\pi/4$
10. 100°
11. $\pi/4$
12. 30°
13. 150°
14. 52.9°
15. $\sin\theta = 1/3$

    $\cos\theta = \sqrt{8}/3$

    $\tan\theta = 1/\sqrt{8}$
16. $\sin\theta = 2\sqrt{2}/3$

    $\cos\theta = 1/3$

    $\tan\theta = 2\sqrt{2}$
17. $\sin\theta = \sqrt{b^2 - a^2}/b$

    $\cos\theta = a/b$

    $\tan\theta = \sqrt{b^2 - a^2}/a$
18. $\sin\theta = y/x$

    $\cos\theta = \sqrt{x^2 - y^2}/x$

    $\tan\theta = y/\sqrt{x^2 - y^2}$
19. $\sin\theta = \sqrt{B^2 - C^2}/B$

    $\cos\theta = C/B$

    $\tan\theta = \sqrt{B^2 - C^2}/C$
20. $\sin\theta = \sqrt{x^2 - r^2}/x$

    $\cos\theta = r/x$

    $\tan\theta = \sqrt{x^2 - r^2}/r$
21. $\sin(-30°) = -\sin 30°$
22. 48°
23. $\cos(7\pi/6) = -\cos\pi/6$
24. $\sin(-45°) = -\sin 45°$
25. $5\pi/12$
26. $\pi/6$
27. $\cos 150° = -\cos 30°$
28. 70°
29. $\sin(5\pi/4) = -\sin(\pi/4)$

Chapter Eleven (cont'd)

30. $\cos(-\pi/2) = +\cos \pi/2 = 0$

31.

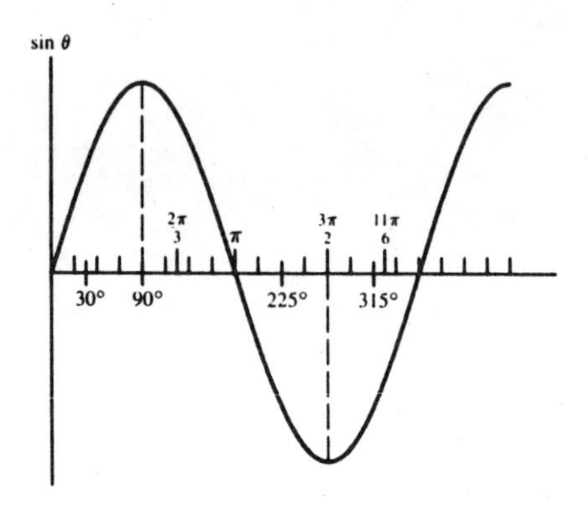

32.

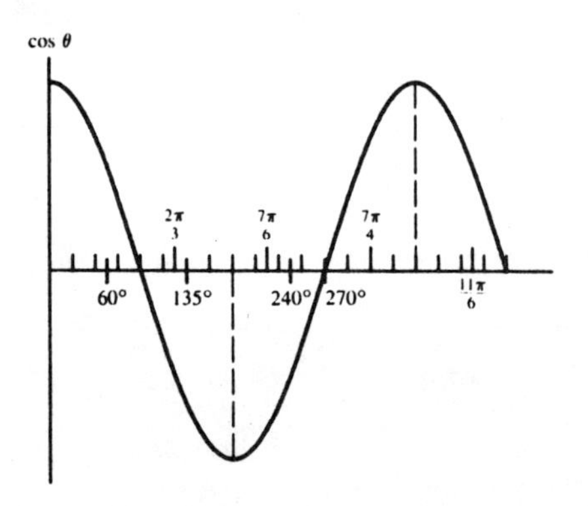

33. $-\sin 40°$

34. $-\cos 60°$

35. $-\tan 47°$

36. $\sin 60°$

37. $-\tan 60°$

38. $-\cos(\pi/6)$

Chapter Eleven (cont'd)

39. tan (π/4)

40. sin (π/3)

41. -cos 60°

42. 30°

43. 75°

44. 89°

45. 85°

46. 150°

47. 105°

48. 402° or 42°

49. 125°

50. 340.5° or 199.5°

51. 116.4° or 243.6°

52. 287.2° or 107.2°

53. 120° or 240°

54. 225°

55. 303.7°

Chapter Twelve

1. 20 x 60 ft

2. 9 years

3. 17 years

4. 6 and 7

5. 16 and 6

6. 280 ft

7. 2.4 hours

8. 3.4 hours

9. 3.3 hours

10. 15 and 35 years